Solvents - Dilute, Dissolve, and Disperse - Insights on Green Solvents and Distillation

Edited by Raffaello Papadakis and Vilmar Steffen

Published in London, United Kingdom

Solvents - Dilute, Dissolve, and Disperse - Insights on Green Solvents and Distillation
http://dx.doi.org/10.5772/intechopen.104128
Edited by Raffaello Papadakis and Vilmar Steffen

Assistant to the Editors: Maqsood Ahmad

Contributors
Ali Boubakri, Amor Hafiane, Anis Bouzeraib, Deepali Jain, Edson Antonio da Silva, Fabio Rodolfo Miguel Batista, Geeta Verma, Jaime Alfonzo Irahola Ferreira, Julio Cesar Ribeiro Nunes, Khaled Basta, Maiquiel Schmidt de Oliveira, Maqsood Ahmad, Mohamed Nadir Khelifi, Muhammad Hammad Rasool, Ouacil Saouli, Parul Tomar, Rafael M. Matricarde Falleiro, Ricardo de Freitas Fernandes Pontes, Salah Al-Tahar Bouguecha, Vilmar Steffen

First published in London, United Kingdom, 2024 by IntechOpen
IntechOpen is the global imprint of INTECHOPEN LIMITED, registered in England and Wales, registration number: 11086078, 167-169 Great Portland Street, London, W1W 5PF, United Kingdom

British Library Cataloguing-in-Publication Data
A catalogue record for this book is available from the British Library

Additional hard and PDF copies can be obtained from orders@intechopen.com

Solvents - Dilute, Dissolve, and Disperse - Insights on Green Solvents and Distillation
Edited by Raffaello Papadakis and Vilmar Steffen
p. cm.
Print ISBN 978-0-85466-137-4
Online ISBN 978-0-85466-136-7
eBook (PDF) ISBN 978-0-85466-138-1

For EU product safety concerns:
IN TECH d.o.o., Prolaz Marije Krucifikse Kozulić 3, 51000 Rijeka, Croatia,
info@intechopen.com or visit our website at intechopen.com.

Meet the editors

Dr. Raffaello Papadakis holds a Ph.D. in physical organic chemistry from the Technical University of Athens, Greece, and a diploma in chemical engineering (MEng). Throughout his career, he has conducted research at various institutions, including CNRS/Aix-Marseille University, France; Uppsala University (Ångström Laboratory and BMC); and as a visiting researcher at CNRS/UBO in Brest, France, the University of Malaga, Spain, and AstraZeneca at Mölndal, Sweden. Previously, he worked as a senior researcher in production and R&D at TdB Labs AB, a manufacturer of polysaccharide derivatives in Uppsala, Sweden. Dr. Papadakis' research interests span physical organic chemistry and materials science, with a specific focus on solvent effects and the chemistry of polysaccharides and graphene. Dr. Papadakis has authored over 40 peer-reviewed scientific papers and book chapters. Currently, he serves as a researcher at the Department of Forest Biomaterials and Technology at Swedish University of Agricultural Sciences, focusing on wood chemistry and adhesives.

Professor Vilmar Steffen received a BSc in chemical engineering from the Western Paraná State University–Unioeste (2007), an MSc in chemical engineering from the Western Paraná State University–Unioeste (2010), and a Ph.D. in chemical engineering from the State University of Maringá (2014). He has successfully completed six months of a postdoctoral research fellowship at Western Paraná State University–Unioeste (2015) and is currently in another postdoctoral stage. Since February 2015, he has been an assistant professor at the Academic Department of Engineering (DAENG) of the Federal Technological University of Paraná (UTFPR) Câmpus Francisco Beltrão, Paraná, Brazil. His research work focuses on process modeling, simulation, and optimization. Recently he has started to work on the proposal of systematic literature review methods and already has five papers related to this subject.

Contents

Preface **IX**

Section 1
Green Solvents and Their Applications **1**

Chapter 1 **3**
Green Solvents
by Parul Tomar and Deepali Jain

Chapter 2 **23**
Green Solvents in Organic Synthesis
by Geeta Verma

Chapter 3 **39**
Deep Eutectic Solvents as a New Frontier in Drilling Fluid Design: Opportunities and Challenges
by Muhammad Hammad Rasool and Maqsood Ahmad

Chapter 4 **59**
1D Modeling of CO_2 Absorption Using K_2CO_3 in a Hollow Fibre Membrane Contactor
by Mohamed Nadir Khelifi, Ouacil Saouli, Anis Bouzeraib and Khaled Basta

Section 2
New Insights on Distillation **71**

Chapter 5 **73**
A Systematic Review of the Literature on Steady-State Reactive Distillation Modeling and Simulation: Challenges and Opportunities
by Vilmar Steffen, Maiquiel Schmidt de Oliveira and Edson Antonio da Silva

Chapter 6 **111**
Evaluation of the Best Operating Conditions in Distillation Columns: A Case Study for the Separation between Nonylphenol and Dinonylphenol
by Julio Cesar Ribeiro Nunes, Ricardo de Freitas Fernandes Pontes, Fabio Rodolfo Miguel Batista and Rafael M. Matricarde Falleiro

Chapter 7 129
Membrane Distillation Process: Fundamentals, Applications, and Challenges
by Ali Boubakri, Salah Al-Tahar Bouguecha and Amor Hafiane

Chapter 8 147
Design, Simulation, and Comparative Analysis of a Carbonating Tower
by Jaime Alfonzo Irahola Ferreira

Preface

In the evolving landscape of chemical engineering and environmental science, the pursuit of sustainability has become a principal concern. This book, dedicated to the exploration of green solvents and distillation, stands at the forefront of this important milestone. It is a collection of recent advances in these two research topics of high scientific and technological importance.

Green solvents have emerged as a groundbreaking alternative to classic solvents since they exhibit low toxicity, they are more environmentally friendly, and they are capable of dissolving a wide range of solutes. Their development is not merely an academic exercise but a necessary response to global environmental concerns. The recent surge in research has revealed novel solvents like supercritical fluids, ionic liquids, and deep eutectic solvents, which promise a reduction in toxic waste and energy consumption. These advancements assist toward a future where chemistry and engineering align with the need for sustainability.

Distillation, the widely known industrial separation method, has also undergone a renaissance. Reactive distillation, a technique that marries chemical reactions with separation, has shown potential for reduced operating costs, improved energy efficiency, and a smaller environmental footprint. Moreover, extractive distillation has recently exhibited significant advances, particularly in the separation of azeotropic mixtures, a long-standing challenge in the field. These innovations in distillation are reshaping the industry, making processes more efficient and less harmful to the environment.

Acknowledging the new trends in these two important fields of research and technology, this book delves into the molecular intricacies of green solvents and their applications and at the same time explores the new trends and technologies in distillation. We are very grateful to all the authors who have contributed chapters to this book, rendering this endeavor possible. We hope that the book will be a useful tool for a wide range of scientists and engineers.

Dr. Raffaello Papadakis
Swedish University of Agricultural Sciences,
Department of Forest Biomaterials and Technology (SBT),
Uppsala, Sweden

Dr. Vilmar Steffen
Academic Departments of Engineering (DAENG),
Federal University of Technology – Parana (UTFPR),
Paraná, Brazil

Section 1

Green Solvents and Their Applications

Chapter 1

Green Solvents

Parul Tomar and Deepali Jain

Abstract

Prolonged use of solvents has harmful impacts on the environment, and entire surviving things. One of the most crucial demands of green chemistry is the cutting down of solvent use or substituting with less hazardous ones. Green solvents are explored as an alternative to traditional solvents. They are specified by low toxicity, easy availability re-usability extraordinary efficiency, environmentally friendly reaction medium, non-flammability, increased reactions rate, decrease reaction temperatures as well greater selectivity, and non-volatility. There are numerous applications of green solvents in many industries such as the cosmetic, pharmaceutical industries, chemical industries, perfumery, and also expended to extract crude material, flavonoids, oils and fats, scents and antioxidants, proteins, volatile compounds, and sugars. To get over the properly-hooked up drawbacks of traditional solvents, exquisite extraordinary research attempts have been these days dedicated to the alternative of conventional chemical reaction media using the so-known as green Solvents. In this experience, the selection of a secure, bio-renewable, non-toxic, and reasonably-priced response media is a vital purpose in chemical synthesis. Thus, this special issue on "green solvents" has been aimed to show off a sequence of exhilarating contributions from global researchers within the various sub-regions of chemical synthesis in green solvents.

Keywords: green chemistry, green solvents, ionic liquids (ILs), water, deep eutectic solvents (DESs), supercritical fluids, extraction, catalysis

1. Introduction

Traditional chemical employment imparts to huge quantity of health and environmental hazards. To scale down these hazards and for accomplishing more environmentally friendly outcomes, the research for alternate options has always been the field of concern. Green chemistry is base for potential alternatives accommodating to the future of chemical industrialization. It is the novel springing up an area that to get the property of being sustainable (sustainability) endeavors to exploit at the stage of molecular extent. According to reports, the field exhibits the capability of safe chemical inventions to accomplish economical and environmental aims. At the opening of the 1990s, a close manner, 20 years ago, the conception of green chemistry has been first invented. It has been defined as the intention of chemical substances and procedures to scale down or get rid of the usage and production of dangerous materials. Green chemistry has assigned a set of 12 principles. These principles approach as a directing framework to the synthesis of noble chemicals regarding holistic access towards green technological development with safety, economy, and biodegradability [1–3].

The 12 principles of green chemistry are summed up under [1]:

1. Nonproliferation: Reducing waste before it multiplies is preferable to removing it afterwords it has been produced

2. Atomic efficiency: Total crude substances should be contained in the very last compounds using synthetic techniques.

3. Involving low risk or dangerous compounds Synthesis: Whensoever possible, it is necessary to choose techniques with minimal negative effects on the surroundings and human beings.

4. Formulating secure chemical substances: Chemicals must be made in a way that preserves the reactions' usefulness while reducing their hazard.

5. More risk free solvents and supporters: It is important to avoid using auxiliary substances wherever possible, and when they are utilized, they must be used sparingly.

6. Invention for energy economy: Energy effect requirements on the surroundings must be identified, and if possible, they should be reduced or changed.

7. Utilize sustainable raw materials: As executable, crude substances should be recyclable as opposed to consume.

8. Step-down derivatives: The production of derivatives ought to be derogated as well as nullified, when viable, and demands extra chemicals and may result in waste.

9. Catalyst: Using catalytic chemicals instead of stoichiometric ones can ameliorate reaction efficacy and decrease energy demand.

10. Planning for degeneration: Chemicals applied must be able to smash down after the reaction in less difficult bio-decomposable substances by response and cease remaining in surroundings

11. Real-time evaluation for contamination interference: Evaluation techniques ought be advanced to allow real-time, procedure observation as well as manipulation dangerous materials prior to production.

12. Implicitly sounder chemistry for accidental injury avoidance: materials for compounds synthesis processes need to prefer for reducing whatever type of chemical injury which includes explosions, and fires

Challenges have surfaced in the sector of chemical treatment due to the daily use of massive quantities of hazardous and combustible solvents. More than 20 million tonnes of organic solvent waste wastes are released into the air every year, resulting in wasteful solvent wastage and environmental pollution. Dimethyl sulfoxide (DMSO), dimethyl formamide (DMF), acetone, and aromatic solvents like benzene, toluene, and chlorinated solvents (CHCl) are examples of organic solvents that contribute to ecosystem contamination but are still widely used. All systems found in living beings are adversely affected by persistent interaction with solvents, but the respiratory and neurological systems are particularly vulnerable. Additionally, the use of dangerous solvents is harmful to organs, for example, chloroform and carbon tetrachloride are hepatotoxic. Glycol ethers and chlorine-based solvents are employed when renal damage arises.

Additionally, dealing with particular chemicals, such as diethylene glycol (DEG), petrochemical distillates, and halogen-containing hydrocarbons, can cause kidney tubular necrosis even quickly. The World Health Organization (WHO) reports that a quarter of all current diseases are caused by prolonged exposure to environmental contaminants. The contaminants accumulate to hazardous levels as a result of the release of manmade compounds or an overabundance of naturally occurring substances. High pollution levels threaten human health, reduce the diversity of species, and damage ecological systems. One of the most significant objectives for sustainable chemistry is to apply fewer solvents or to replace them with ones that are less harmful.

The aforementioned information is of tremendous interest to both academics and industry regarding the usage of environmentally conscious solvents [4–10].

2. Traditional solvents and their hazards

Traditional solvents are common laboratory solvents that we use every day for industrial purposes as well as for laboratory because of their melting point, volatility, better dissolution power, and yield. Aside from their addressed influence on lab workers, when liberated in surroundings such as air, soil, and water, these can induce strong threats to plants, human beings, and animals from the bottom level attaining to the bulk level when leftover untreated [11]. Most of the chemicals are dangerous if not managed cautiously. Direct interaction with the skin, breathing, and eye contact and extended, and exposure are the two of greater importance path of chemical perniciousness to mankind. Damaging to health, harm to organs, breaking of the immune system, growth of allergies, respiratory disorders, generative disturbs and congenital abnormality, effects on the mental condition and intellect or physiological growth of children, cancer, etc. can rely on the chemical substance constitution [12].

According to the Team for European Solvents Commercial Enterprise recommendation for following European Union (EU) regulations ameliorate atmosphere attribute via encouraging step-down entire volatile organic compounds (VOCs) release after 1990s [13]. Unyielding chemicals deposition inside surroundings, and fallaciously unwanted materials treatment are base reason for environs contamination for a period and stimulates serious toxicity to the earth, atmosphere, and water body [14]. Ozone layer depletion, land and water pollution, biological oxygen demand (BOD), and chemical oxygen demand (COD) are the most crucial consequences of extensive expenditure on traditional solvents. The Solvents Industry Association provides proposals and instructions to manufacturers, suppliers as well as exploiters of solvents to facilitate derogating potential environmental strikes [15]. Industry and producing corporations usually depend on solvents for a large number of works like manufacturing products, scavenging and getting rid of grease or oil from surfaces and machinery, exploiting chemicals that aid in chemical reactions, like paints, and coatings. In order to reduce the utilization of organic volatile solvents with having negative effects on human health and the environment, various organizations with sustainability programmes are searching for environmental amicable resolutions. Different conventional solvents, their sources, and related risks to human health and the environment.

Utilizing reusable solvents and incineration, two balancing techniques can be used to reformulate energy. Distillate is used in the refinement of used solvents, which uses more energy than producing new solvents to a comparable degree. Energy is directly produced by incinerating waste, although this process necessitates the production of additional solvent locally. The approach that provides the greatest cumulative energy demand (CED) discount is dependent upon the type of solvent.

3. Definition of green solvents

In the headline of their 2007, Capello and Fischer [16] have posed the query at ETH Zurich (in any other case called as a Swiss Federal Institute Technology) is an elementary; "what is a green solvent"? Their reply may be considered like a fast life cycle assessment (LCA) kind estimation and now is prestigious, cuckold environmental evaluation, guard, welfare, and energy requirement. After recognizing the energy demanded to develop solvents, and the alternatives to be had at quit-of-existence to retrieve some of that energy, the ultimate CED for solvent manufacturing can be figured out. Utilizing reusable solvents and incineration, two balancing techniques can be used to reformulate energy. Distillate is habituated in the refinement of applied solvents, which consume more energy than manufacturing latest solvents to a comparable extent. Incinerating wastes straightforwardly results in energy, although this procedure calls for the manufacture of surplus solvents topically. Method which gives more CED discount is reckoned upon the kind of solvents. The energy for distillation of a solvent is lesser than produced. The incineration credit is the energy retrieval from combustion, yielding a discounted CED. According to modified LCA admittance, most (however no longer all) hydrocarbons are great incinerated for example *n*-hexane however no longer toluene, and the identical enforces to diethyl ether. The functionalized solvents with more lifelong manufacturing ways are quality recycled to hold the energy and merit invested into the molecules throughout their authentic synthesis for example DMF, and blessings for ethyl alcohol are rather evenly distributed. The same authors have produced still more precise elaborated valuation of energy call for solvents output [17].

4. Uses, sources, and types of green solvents

Solvents employed in one-of-a-kind chemical procedures are determined to fundamental sustainability challenges [4]. Yearly boom records have reported that mass of chemical manufacturing enterprise through Europe has anticipated for booming by 3.1% showing greater participation, contributing to environmental reduction [18]. Thus, step-down the employment of solvents, and substituting these with low poisonous greener dissolvents are two main goals for sustainable chemistry [19].

Fisher asserts that the greener solvents press out a target for diminishing in the atmospheric effect of solvent intake in chemical manufacturing [20, 21]. Efficient exercise of green solvents consisting of water, ionic liquids (ILs), supercritical fluids, and liquid polymers can assist reap the objective to cut down environmental depletion. Low harmfulness, section conduct, chemicals, thermodynamics, non-flammability, and biodegradation are some of the homes of novel biosolvents or green solvents [22]. And indicates one-of-a-kind sorts of green solvents at the side of their respective examples and discusses unique solvents their sources, commercial, and standard makes use of.

In the separation of terpenes, plant based substances and oils serve as a terrific alternative to petrochemical solvents such as n-hexane [23]. Water, existing as a worldwide greener media and not-poisonous, may be utilized inside numerous response mediums, separation, extraction, and chemical formations [24]. Deep eutectic solvents (DESs) are some other environmentally beneficial green solvent that are extensively applied during diffusive liquid–liquid based microextraction [25], and ecological absorbent materials for risky chemical contaminants [26]. Some other amazing magnificence categories of green solvents admit ILs, that are utilized in commercial settings exploited in getting rid from sulfur from petroleum-based products [27] and carbon absorption [28]. Rather than utilization of *n*-hexane, *n*-butane may be consumed to serve as an attainable solvent during plant-based materials isolation [29]. In processor for polymers melting [30] and prior-remedy of lignin-based biomass [31]. Super-critical liquids are another class of green solvents that can be the best alternative to conventional solvents. Fatty acid esters are a good desired as a solvent components for carotenoid extraction [32]. Bioethanol, due to its low toxicity and constant obtainability, is habituated as a green solvent in numerous synthesizes and the industry manufactures [33]. Because of its renewability, easy decomposition, cost-effective, and eco-friendly, castor oil functions as auspicious green solvent for nanoparticles synthesis [34], agriculture, meals, fabric, papers, plastics, rubber, cosmetics, perfumes, coatings, electronics, pharmaceutics, paints, additives, ink, lubricating substance, and biofuels [35].

Bio-derived specially solvents which include ethyl lactate also are a sustainable solvent in the treatment of green tea biomolecules [36] as well as in the chemical reactions notably and in synthesis [37]. Polyethylene glycol (PEG) is also a green alternative that is widely known for its carbon–carbon bond-making and coupling reactions [38].

5. Synthesis schemes with green solvents

In all product formation procedures and distinct applications in industry, solvents are wanted in large quantities resulting in ample quantities of waste. Path-breaking technologies and distinct synthesis strategies have been discoursed solventless operations that are not admitted for every field of study because of a few marketplace issues. Later on solventless thoughts, chemists as well as medicinal scientists investigated solvents that fit green chemistry. Fischer has reported that green solvents explicitly have the objective to reduce the environmental effect triggered by the intake of chemicals throughout manufacturing [21]. For solvents that are considered greener, certain techniques have been merged. The following are alternatives to dangerous solvents that are extra kinder to the planet, recyclable, and downplaying for ozone layer exhaustion potency, habituation of bio-solvents or oleochemicals, and replacement of organic solvents which are ILs, and supercritical fluids.

According to the literature review, various examples of green solvents can be visible for cooking natural products, drugs, and crucial intermediate productions which can be applied to foster synthesis [16].

6. Ionic liquids as green solvents

Magical chemical substance ILs, having precisely tunable houses with loads of packages applications in each area have arrested brilliant attraction from scientists universally. The history of liquids efficiently commenced in 1914 while the beginning of the early IL ethyl ammonium nitrate [$EtNH_3$] [NO_3], having a melting state of 12°C has been reported. Later on, a progression of first-rate associated ILs have been exposed remarkably, have demonstrated tremendous pursuit within the chemicals industries as well as in academical due to their amazing greener dimensions and pose as a liquid at beneath a 100°C temperature. ILs do not have any borderline between inorganic and organic compounds due to their tunable characteristics. Additionally, goal-specified ILs with adjustable properties can be fashioned using converting cation-anion combining. Examples of various ordinarily devoted cationic parts are pyridinium [39], imidazolium [40], ammonium [41], pyrrolidinium [42], phosphonium [43], and piperidinium [44] with version constituents and anionic substituents tetrafluoroborates, alkyl sulphonates, alkyl tosylates, and hexafluorophosphates for manufacturing of ILs. As an end outcome, ILs have very much blessings over conventional organic solvents consisting of magnificently down volatility, worthless vapor strain [45], less dangerous, non-flammability [46], dissolubility, and greater.

Environment compatible, viscosity [47], excessive chemical as well as thermal stability [48], and eminent ionic conductance [49–52]. Ordinarily, the viscosity of ILs swears by and complements with growing hydrogen bondings, the forces produced by van der Waals, hydrocarbon chains dimensions, along with the volume comprising fluorinations and anions. Moreover, ILs usually have a viscosity that fluctuates between 10 and 500 mPa at ambient temperature [53–56].

7. Ionic liquids as ionic salts

ILs are tremendously wonderful and different from ionic salts like NaCl [57]. Ionic salts lie in a solid state at room temperature and also have electrostatics interplay with generally metal cations and non-metal anions.

Wholely, ionic substances inclusive of NaCl can be in a molten or liquid phase at excessive temperatures. Therefore, ionic salts are incapable of being applied to the establishment of organic compounds. Where, ILs, a class of ionic salts with a melting state beneath 100°C, can be beneficial for chemical synthesis due to their ability to dissolve in inorganic, organometallic, and organic materials [58].

8. Water as green solvent

Water exhibits numerous peculiar physicals and chemical homes including massive hydrogen bonding, excessive heat potentiality, big dielectric steady, and a big temperature range. Water as a universal solvent has consequently lots of benefits over traditional solvents. Moreover water can be decided on as a green solvent because of its cost-effectiveness, no trouble to be handled, safe, no risk, nonpolluting, and easy as well as quickly obtainable, and inflammable. As a matter of fact, human beings do not address water to be known as a chemical. Despite a large number of crucial benefits of water, it's far still no longer usually applied to be known as a sole solvent for synthesis schemes in laboratories and industries as the maximum of the chemical components do not have the property of being dissoluble throughout water. Since water has been the medium of preference for our planet's biochemical processes for hundreds of years, researchers are striving to replicate the inherent synthesis procedures that occur in water. As a consequence of an antique scientific hypothesis that claims irreducibly undissolved chemicals can not anymore produce any kind of substance, investigators found themselves off out of the water during a significant amount of time. Simultaneously, Sharpless has changed such superannuated concept by branded novel wondering thoughts such responses must be advanced "along" or "inside" water, by proposing that dissolvability is not vital to chemical responses. Sharpless has discovered procedures inclusive of cycloadditions, nucleophilic establishing of epoxides, Diels-Alders, and Claisen rearrangements in that as the reactants had been not dissolved in water, the reactions were described as being on-water [59].

Despise the fact that water reasons many issues as a solvent in chemical synthesis, purification processes, and separation of end products, Recently, it has been turned very much preferred in organic reactions. In a few instances, water accelerates reactions by dissolving the reagents and guarantees selectivity. Furthermore, oxygen can not be dissolved in water which may be a bonus for metallic catalysts. In the closing decade water has grown to be an actually beloved solvent encourages of its characteristics which include being nonhazardous for the natural world and has been authenticated in a bombastic quantity of papers [60–62]. It has been acknowledged that water is the most addictive solvent in every biochemistry reaction conveyed in nature [61]. Nevertheless, water is an appropriate solvent only for organic reactions, having polar groups such as carboxylic acids or alcohols. Notwithstanding, two centuries ago unique selectivity existed in aqueous solvents, and was additionally located in water suspensions, when one dissoluble component interacted with a badly dissoluble one [5, 61, 63]. Recently, various scientists have discovered an awesome speedup of chemical processes in this sort of suspension, assigned by them: reactions on water. It could be a statement to mention that water is exactly an eco-friendly solvent, it dominates notable homes which are substantially precise, related to the hydrophobic phenomenon [59, 61, 63, 64].

9. Deep eutectic solvents as green solvents

Firstly Abbott et al. have defined DES [65] in the context of a sustainable solvent-based combination of choline chloride together with urea an example of, whilst combined in an adequate mole ratio, demonstrated meltdown factors lots below compared to the values of any one particular part of many constituents. The term 'eutectic' arises from the Greek word ευτηκτος, interpreted as easily melting', and in 1884, British investigator Guthrie [66] utilized it for elucidating a lesser degree for the liquefying compared to specified by existing proportionality. In the future, it becomes crucial to keep in mind, nevertheless, in which just a few eutectic combinations deserve to be designated as DES since, hardly every combo of substances which remain insoluble inside a stable segment necessarily demonstrate an eutectic state, as pointed out throughout the discussion through Martins et al. [67]. According to International Journal of Molecular Science, 2022, for the qualifying word 'deep', to date, there is no cosmically harmonized clarification, however maximum of the litterateurs coined the word 'deep' as those mixtures with a eutectic temperature some distance beneath that of a model liquid solution [68–70]. Even though distinct appellations are exploited, the precept implemented to categorize DES is identical and the DES acronym is still utilized in an always globalizing idea. Smith et al. [71] recalled the word 'deep eutectic' to encompass the eutectic combination of Brønsted or Lewis acids as well as bases which contain more than a few cationic and anionic species and categorized them into four distinct kinds because of their materials [71]. Particularly, DES can be specified by an arrangement of one or more hydrogen bonds donator (HBD) as well as hydrogens bonds admittance chemical substances those, with a proper mole proportions, display an eutectic state temperatures under those for an outstanding solutions [67–70].

10. Peculiar physicochemical characteristics of deep eutectic solvents

The fascinating consequence demonstrated through DES is that solid pure constituents can get liquid via intermixing them in a sure proportion underneath a moderate heating. An elaboration of this upshot in DES formulation is shown, displaying that sugars like fructose, sucrose, glucose, and malic acid are solid at room temperature due to it has melting state over 130°C, at the same time as their combos (denoted as DESs) are in liquid shape [72]. On blending, the HBD and hydrogen bonds acceptor (HBA) constituents will intercommunicate with one another by hydrogen bonding. A naive example of DES synthesis made up of choline chloride and HBD via hydrogen bonding interactions is displayed. As proven, there's a formation of hydrogen bonding interactions among the chloride ion of HBA with the OH groups of HBD [73].

11. Vegetable oils as green solvents

Vegetable oils, are oleochemicals that are pressed out from various plant sources and are renewable resources and have triglyceride shape. Vegetable oils are made from glycerol by the substitution of all hydroxyl groups of glycerol with various fatty acids which develop them into solid or liquid productions [74]. Vegetable oils are crucial meals constituent. Unluckily, they have now not been taken into consideration as a green solvent so far besides for some reactions such as the acylation reaction in corn oil [75]. Vegetable oils have been applied for biopolymers and are probably accepted by researchers who are seeking out brand new sources of green solvents. The cyclization and acylation reactions have occurred in vegetable oils, particularly corn oil. Applications and characteristic properties of vegetable oils have been discoursed and yields, time of reaction, and sustainability of vegetable oils have been equated with each other as well as with hazardous solvents such as xylene. A combination of dibenzoylmethane, phenol, and oxalyl chloride was heated in the presence of corn oil at 120°C for 15 minutes. Scientists have interpreted that CH_2 of dibenzoylmethane turned into acylated very effortlessly. This reaction is the primary instance of vegetable oils and this concept must be focused on utilizing greater synthesis strategies because of the cost and effectiveness of vegetable oils [75].

12. Supercritical water and carbon dioxide as green solvents

The supercritical fluids, as the name suggest, are chemicals that reside in an intermediate phase across a substance that is liquid along with an air at pressures and temperatures over the corresponding critical points [60, 76]. The supercritical fluids may work such as gases or liquids and can be bypassed through solid substances or liquids like beverages, dissolving other substances. Furthermore, when each pressure and temperature are close to the critical level, converting them even marginally can bring about grievous density alterations.

Supercritical fluids are an aptest alternative to organic solvents for commercial and laboratory procedures. Generally, supercritical carbon dioxide (sCO_2) or water is ordinarily expended supercritical fluid and is taken into consideration to greener solvent [60]. Owing to the extremely good dissolubility inside lots of polymer, supercritical carbon dioxide (sCO_2) has an unreplaceable issue for polymers procedure. Moreover, the selection of less dangerous fluids like supercritical carbon dioxide (sCO_2) in place of traditional organic solvents is a bonus to the surroundings [60, 77].

13. Applications of green solvents

ILs are utilized in the desulfurization technique to diminish SO2 eliminations in the petroleum enterprise [78]. The key industrious applications of green solvents have a huge variety from extraction processing to chemical synthesis such as sensors as well as biosensors, CO2 captured, lignocellulose biomass use, and bio-supported chemical substances [79]. Solvents utilized inside the pharmaceutic discipline reports for mass intake inside the variety of 80–90% and biosolvents for example butanol, glycol, polyethylene, bio-reusable materials along with supercritical fluids dissolvents have been extensively employed inside recovery and in the dissolution to drug [80]. The most aggressively addressed sustainable solvents during the aforementioned period were supercritical liquids, ILs along with deep eutectic fluids, predominantly used for the treatment of herbal remedies, foodstuffs, flavor, including perfume, in addition to the processor of medicine plants [81]. Organic carbonates are a category of organic compounds and reaction intermediates that are recognized for their eminent bio-degradation, low harmfulness, and versatility which cause their extensive applications in bio-catalysis [82]. Unconventional solvents are used for biomass pre-treatment as well as separation of natural polymeric materials from assets let in ILs, bio-based solvents, and DESs [83]. Another fascinating field is rare earth elements detail restoration and ionic solvents, which include ILs and deep-eutectic solvents, which are becoming a variety of hobby due to the fact they provide an alternate options to classical metallic healing techniques [84].

14. Green solvents in chemical reactions

With the developing advancements within the discipline of chemical studies, the capacity for danger to surroundings and human health has accelerated appreciably. Thus, to establish a chemist, extra surroundings-friendly magnificent interest has been paid to the characteristics and usages of green solvents in chemical processes, extractions and fractional processes, and synthesis of materials. Lots of procedures that exercise green solvents have been used for commercial purposes [85]. More a couple of examples of organic synthesis are addressed hereby utilizing green solvents as a replacement.

1. Glycerol can work like a reducing agent in the presence of a metal catalyst in reactions, switch the hydrogenation of olefins, and might yield satisfactory outcomes and can be reused later on [85].

2. In the Suzuki chemical reaction, the yield was nice among 4-iodotoluene and phenylboronic acid in PEG400 (90%) in comparison to other solvents including DMF and toluene.

3. Ethylene glycol is a plentiful recyclable biomass, low-toxicity of solvent, that has prominent industrialized significance and is an outstanding green solvent for photo-catalyzed cascade cyclization processes [86].

15. Green solvents in environmental acceptable extraction techniques

Excellent belongings of ILs, potentiality to synthesize with specific dissolubility in water, and in organic mixtures like ethanenitrile or methanol causes their applications of extractions excitingly. Because of their viscosity, ILs have been applied in the chromatographic techniques as additives and after including high-performance liquid chromatography solutions, viscosity of ILs minimized [87, 88]. In unique extraction techniques capillary electrophoresis (CE), ILs are used as heritage electrolytes and additives. In Addition, ILs qualify the capillaries walls as well as electrophoresis separations [89–91]. ILs are also utilized inside the extraction of flavonoids compound from phenolic chemicals and imidazolium supported ILs, and ketoprofen, carprofen, naproxen, and suprofen by the ILs (C_4mim) (NTf_2). Researchers of natural products use applying of ILs as green solvents to isolate and extract from plants such as terpenoids, alkaloids, flavonoids, and phenolics compounds [92–95].

Traditionally, investigators on the whole make use of unsafe unstable organic compounds (VOCs) in extraction techniques to have chemicals extracted like maceration soxhlet, distillations, percolations, and infusions. However, all chemists have long tried for substituting classical techniques that have an environment strike with green techniques such as microwave assisted extractions, supercritical fluids extractions, ultrasound-assisted separations, excessive-rate homogenizations, and pulsed electric field, and pressurized solvents. ILs scale back solvent uptake and extraction time. Supercritical fluids extraction has amazeballs extraction houses but a huge operating cost than ILs [96–100].

16. Green solvents as a catalyst for the preparation of bio-diesel

Worthless vapor pressure, excessive thermal constancy, state transformations conduct, dissolubility and mixability with chemicals, basic, and acidic, swappable ILs as green solvents, biodiesel production by ILs, catalyst, solvent for enzymes-catalyzed transesterifications, catalyst based, reusing for ILs, restoration of ILs, DESs which includes choline chlorides supported, the weird family for ILs seemed such a noble era for ILs which shape hydrogens bondings. In addition, hydrogens-bondings interplay in DES are high-energy advantageous as an outcome of greater dissolubility. Scientists have habituated a combining of ILs with deep eutectic solvents in biodiesel practise [101, 102].

17. Future approaches

Green chemistry has the potential to be leveraged to accomplish sustainability across three critical domains. Initially infinite inexhaustible energy resources, whereby scientists foster the financially feasible process for converting the sun's energy towards chemically generated energy. Secondly, petrochemicals are the biggest supplier of the chemicals utilized throughout the manufacturing industries, along with more and more need to be sourced from renewable resources to subdue our habituation of fossil carbon. Thirdly, technologies that might be harmful to the surroundings need to get replaced by way of nontoxic alterations.

The biannual environmentally conscious solvents convention, which continues to conduct events during the time 2002, has grown into a crucial venue for conversation to breakthroughs in science including business application by using ameliorated solvents enabling reactions processes as well as approaches. These multidisciplinary assembly dealt the exploitation with applications of the optional solvents, admitting aqueous stage to inappropriate applications, ILs and supercritical liquids, in addition to green natural solvents and soluble polymers. Other subjects have been admitted protected without solvents procedures, and substances treating solutions, phases extractable chemicals, with numerous abstraction techniques [103]. This extraction performance and compounds structures fit nicely to industries functions.

Followings include some alternates for common solvents:

- 2-MeTHF as well as CPME both are created by furfural, that are the biology-based compounds. These are extremely helpful to substitute dangerous chemical compounds inclusive of THF. 2MeTHF an outstanding benefits for demanding natural biological cycles. Solvents close off circle are much critical for making sure, not again harm for surroundings takes place. These are splendid opportunities to have a look at the applicability of 2MeTHF within the chemical processing industries [104].

- Propylene carbonate (PC) is also an essential green solvent with a purpose to eat CO2 that is provided. Several biological operations can develop this CO2, such as bio-ethanol. PC has been also discovered as a desirable replacement to THF.

- The beauty sector has made 1,3-propanediol a crucial solvent, and it can be applied to an extensive variety of applications along with ketonic group inside chemical solvents. Additionally, it can be deployed in the creation of recyclable polymers [105].

- DCM solvent has been upgraded with ethyl lactate. Additionally, it has applications of excellent potency to eliminate caffeine of tea [106].

18. Conclusions

Our forthcoming natural world sustainability challenges demand an effective alternatives, and green technologies for paving the manner for such an approaches. Glycerol, ILs, water can be taken into consideration as the naive green solvents as well as their overall functioning and top production, these include the most frequently employed commercial noble glowing solvents of choice, as demonstrated by outcomes. Additionally other solvents, such as 2-MeTHF, CPME, and ethyl lactate, have been verified as more environmentally friendly alternatives to standard solvents in terms of inhibiting and avoiding contamination. Twelve ideas of sustainable chemistry shape an idea of developing greener solvents inside gases phase, liquids phase, and solid phase with appropriate efficiency. According to European solvent industrial enterprise, the complete ban of VOCs will damage the industry but their proper and acceptable management is essential to scale down the solvent-supported whole environment depletion.

Acknowledgements

I owe my profoundest admiration to my supervisor, Assoc. Prof. Deepali Jain of D.N. College, Meerut for unforgettable affirms, directions, advancement, and freedom to investigate novel viewpoints in my field of study. She, the inspiration behind these paintings, created a partner for me in the realm of green chemistry and green solvents.

She supported me through all of the issues we encountered while serving as my supervisor, allowing me to take on new tasks. I want to express my gratitude to the entire faculty of the chemistry department at N.A.S. college, CCS University, Meerut, as well as to all of my colleagues for their support and for providing me with tonnes of inspiration and ideas.

Competing interests

The authors say they have no competing interests.

Author details

Parul Tomar[1*] and Deepali Jain[2]

1 Department of Chemistry, N.A.S. College, CCS University, Meerut, Uttar Pradesh, India

2 Department of Chemistry, D. N College, CCS University, Meerut, Uttar Pradesh, India

*Address all correspondence to: parultmr@gmail.com

References

[1] Anastas P, Eghbali N. Green chemistry: principles and practice. Chemical Society Reviews. 2009;**39**(1):301-312. DOI: 10.1039/b918763b

[2] Anastas PT, Williamson TC. Green Chemistry: Designing Chemistry for the Environment. Washington, DC: American Chemical Society, ACS Symposium Series No 626; 1996

[3] Collins TJ. Green chemistry. In: Lagowsky JJ editor. Macmillan Encyclopedia of Chemistry Simon and Schuster. Vol. 2. New York: Macmillan; 1997. pp. 691-697

[4] Welton T. Solvents and sustainable chemistry. Proceedings of the Royal Society. 2015;**471**(A):20150502

[5] Jutz F, Adanson JM, Balker A. Ionic liquids and dense carbon dioxide: a beneficial biphasic system for catalysis. Chemical Reviews. 2011;**111**(2):322-353

[6] Sanni Babu N, Mutta Reddy S. Impact of solvents leading to environmental pollution. National Seminar on Impact of Toxic Metals, Minerals and Solvents leading to Environmental Pollution. Journal of Chemical and Pharmaceutical Sciences. 2014. ISSN: 0974-2115 (Special Issue 3: October 2014)

[7] Dick FD. Solvent neurotoxicity. Occupational and Environmental Medicine. 2006;**63**(3):221-226

[8] Malaguarnera G, Cataudella E, Giordano M, Nunnari G, Chisari G, Malaguarnera M. Toxic hepatitis in occupational exposure to solvents. World Journal of Gastroenterology. 2012;**18**(22):2756-2766

[9] Lauwerys R, Bernard A, Viau C, Buchet JP. Kidney disorders and hematotoxicity from organic solvent exposure. Scandinavian Journal of Work, Environment & Health. 1985;**11**(1):83-90

[10] Rama Koteswararao P, Tulasi SL, Pavani Y. Impact of solvents on environmentl pollution. National Seminar on Impact of Toxic Metals, Minerals and Solvents leading to Environmental Pollution. Journal of Chemical and Pharmaceutical Sciences. 2014. ISSN: 0974-2115 (Special Issue 3: October 2014)

[11] Available from: https://www.cdc.gov/niosh/topics/emres/chemagent.html [Last accessed: 02 Oct 2021]

[12] Available from: https://www.canada.ca/en/health-canada/services/health-effects-chemical-exposure.html [Last accessed: 30 Sep 2021]

[13] Available from: https://www.esig.org/regulatory/air-quality [Last accessed: 25 Sep 2021]

[14] Available from: https://www.environmentalpollutioncenters.org/organic-solvents [Last accessed: 25 Sep 2021]

[15] Available from: https://www.solvents.org.uk/solvents-and-the-environment [Last accessed: 02 Oct 2021]

[16] Capello C, Fischer U, Hungerbuhler K. What is a green solvent? A comprehensive framework for the environmental assessment of solvents. Green Chemistry. 2007;**9**:927-934

[17] Capello C, Wernet G, Sutter J, Hellweg S, Hungerbuhler K.

A comprehensive environmental assessment of petrochemical solvent production. International Journal of Life Cycle Assessment. 2009;**14**:467-479

[18] Available from: https://www.statista.com/statistics/407861/forecast-for-annual-growth-in-chemical-industry-worldwide-by-region [Last accessed: 23 Sep 2021]

[19] Agata T. Green solvents. Journal of Education, Health and Sport. 2017;7(9):224-232

[20] Menges N. The role of green solvents and catalysts at the future of drug design and of synthesis. In: Saleh HEDM, Koller M, editors. Green Chemistry. Boston, USA: IntechOpen; 2017. pp. 74-89. Ch. 5

[21] Simon MO, Li CJ. Green chemistry oriented organic synthesis in water. Chemical Society Reviews. 2012;**41**(4):1415-1427

[22] Greer AJ, Jacquemin J, Hardacre C. Industrial Applications of Ionic Liquids. Basel Switzerland: Multidisciplinary Digital Publishing Institute; 2020. pp. 1-2

[23] Tanzi CD, Vian MA, Ginies C, Elmaataoui M, Chemat F. Terpenes as green solvents for extraction of oil from microalgae. Molecules Journal. 2012;**17**(7):8196-8205

[24] Breslow R. The principles of and reasons for using water as a solvent for green chemistry, part 5 reactions in water. In: Li CJ editor. Handbook of Green Chemistry. Germany: Wiley-VCH Verlag GmbH & Co. KGaA; 2015. pp. 1-3

[25] Wanga XL, Lub Y, Shib L, Yanga D, Yanga Y. Novel low viscous hydrophobic deep eutectic solvents liquid-liquid microextraction combined with acid base induction for the determination of phthalate esters in the packed milk samples. Microchemical Journal. 2020;**159**:105332

[26] Moura L, Moufawad T, Ferreira M, Bricout H, Tilloy S, Monflier E, et al. Deep eutectic solvents as green absorbents of volatile organic pollutants. Environmental Chemistry Letters. 2017;**1**:2-5

[27] Abro R, Abdeltawab AA, Al-Deyab SS, Yu G, Qazi AB, Gaoa S, et al. A review of extractive desulfurization of fuel oils using ionic liquids. Royal Society of Chemicals. 2014;**67**:5-6

[28] Shukla SK, Khokarale SG, Bui TQ, Mikkola JT. Ionic liquids: potential materials for carbon dioxide capture and utilizations. Frontiers in Materials. 2018;**6**:42

[29] Rapinel V, Rombaut N, Rakotomanomana N, Vallageas A, Cravotto G, Chemat F. An original approach for lipophilic natural products extraction: use of liquefied nbutane as alternative solvent to n-hexane. LWT-Food Science and Technology. 2016;**85**:524-533

[30] Kemmere MF, Meyer T. Super critical carbon dioxide. In: Meyer T editor. Polymer Reaction Engineering. Germany: Wiley-VCH Verlag GmbH & Co. KGaA; 2005. pp. 3-8. Ch. 5

[31] Escobar ELN, da Silva TA, Pirich CL, Corazza ML, Ramos LP. Supercritical fluids: a promising technique for biomass pretreatment and fractionation. Frontiers in Bioenjineering and Biotenchnology. 2020;**8**:252

[32] Diacon A, Calinescu I, Vinatoru M, Chipurici P, Vlaicu A, Boscornea AC, et al. Fatty acid ethyl esters (FAEE): a new, green and renewable solvent for the extraction of carotenoids from tomato waste products. Molecules. 2021;**26**:4388

[33] Available from: https://www.merckmillipore.com/IN/en/reagents-chemicals-and labware/bioethanol/86Kb.qB.LIk AAAFDdo9xDMUX,nav?ReferrerURL=https%3A%2F%2 Fwww.google.com%2F [Last accessed: 27 Sep 2021]

[34] Mensah MB, Awudza JAM, 'O'Brien P. Castor oil: a suitable green source of capping agent for nanoparticle syntheses and facile surface functionalization. Royal Society Open Science. 2018;**5**:180824

[35] Mubofu EB. Castor oil as a potential renewable resource for the production of functional materials. Sustainable Chemical Processes. 2016;**4**:11

[36] Villanueva-Bermejo D, Reglero G, Fornari T. Recent advances in the processing of green tea biomolecules using ethyl lactate a review. Trends in Food Science and Technology. 2017;**62**:1-12

[37] Paul S, Pradhan K, Das AR. Ethyl lactate as a green solvent: a promising bio-compatible media for organic synthesis. Current Green Chemistry. 2016;**3**(1):112-117

[38] Vafaeezadeh M, Hashemi MM. Polyethylene glycol (PEG) as a green solvent for carbon-carbon bond formation reactions. Journal of Molecular Liquids. 2015;**207**:73-79

[39] Pont AL, Marcilla R, De Meatza I, Grande H, Mecerreyes D. Pyrrolidinium-based polymeric ionic liquids as mechanically and electrochemically stable polymer electrolytes. Journal of Power Sources. 2009;**188**(2):558-563

[40] Freire MG, Neves CM, Marrucho IM, Coutinho JA, Fernandes AM. Hydrolysis of tetrafluoroborate and hexafluorophosphate counter ions in imidazolium-based ionic liquids. The Journal of Physical Chemistry A. 2009;**114**(11):3744-3749

[41] Kogelnig D, Stojanovic A, Galanski M, Groessl M, Jirsa F, Krachler R, et al. Greener synthesis of new ammonium ionic liquids and their potential as extracting agents. Tetrahedron Letters. 2008;**49**(17):2782-2785

[42] Gao H, Luo M, Xing J, Wu Y, Li Y, Li W, et al. Desulfurization of fuel by extraction with pyridinium-based ionic liquids. Industrial & Engineering Chemistry Research. 2008;**47**(21):8384-8388. DOI: 10.1021/ie800739w

[43] Breitbach ZS, Armstrong DW. Characterization of phosphonium ionic liquids through a linear solvation energy relationship and their use as GLC stationary phases. Analytical and Bioanalytical Chemistry. 2008;**390**(6):1605-1617

[44] Paduszyński K, Domańska U. Experimental and theoretical study on infinite dilution activity coefficients of various solutes in piperidinium ionic liquids. The Journal of Chemical Thermodynamics. 2013;**60**:169-178

[45] Earle MJ, Esperança JM, Gilea MA, Lopes JNC, Rebelo LP, Magee JW, et al. The distillation and volatility of ionic liquids. Nature. 2006;**439**(7078):831-834

[46] Ye C, Liu W, Chen Y, Yu L. Room-temperature ionic liquids: a novel versatile lubricant. Chemical Communications. 2001;**21**:2244-2245

[47] Seddon KR. Ionic liquids for clean technology. Journal of Chemical Technology and Biotechnology. 1997;**68**(4):351-356

[48] He Z, Alexandridis P. Nanoparticles in ionic liquids: Interactions and organization. Physical Chemistry Chemical Physics. 2015;**17**(28):18238-18261

[49] Noda A, Hayamizu K, Watanabe M. Pulsed-gradient spin-echo 1H and 19F NMR ionic diffusion coefficient, viscosity, and ionic conductivity of non-chloroaluminate room temperature ionic liquids. The Journal of Physical Chemistry B. 2001;**105**(20):4603-4610

[50] Welton T. Room-temperature ionic liquids. Solvents for synthesis and catalysis. Chemical Reviews. 1999;**99**:2071

[51] Shariati A, Gutkowski K, Peters CJ. Comparison of the phase behavior of some selected binary systems with ionic liquids. AIChE Journal. 2005;**51**:1532

[52] Brennecke JF, Maginn EJ. Ionic liquids: Innovative fluids for chemical processing. AIChE Journal. 2001;**47**:2384

[53] Varma R. Solvent-free organic syntheses. Using supported reagents and microwave irradiation. Green Chemistry. 1999;**1**(1):43-55

[54] Davoodnia A, Heravi MM, Safavi-Rad Z, Tavakoli-Hoseini N. Green, one-pot, solvent free synthesis of 1, 2, 4, 5-tetrasubstituted imidazoles using a Brønsted acidic ionic liquid as novel and reusable catalyst. Synthetic Communications. 2010;**40**(17):2588-2597

[55] Endres F, ElAbedin SZ. Air and water stable ionic liquids in physical chemistry. Physical Chemistry Chemical Physics. 2006;**8**(18):2101-2116

[56] Bonhote P, Dias A-P, Papageorgiou N, Kalyanasundaram K, Grätzel M. Hydrophobic, highly conductive ambient-temperature molten salts. Inorganic Chemistry. 1996;**35**(5):1168-1178

[57] Dupont J. From molten salts to ionic liquids: a "nano" journey. Accounts of Chemical Research. 2011;**44**(11):1223-1231

[58] Hough-Troutman WL, Smiglak M, Griffin S, Reichert WM, Mirska I, Jodynis-Liebert J, et al. Ionic liquids with dual biological function: sweet and antimicrobial, hydrophobic quaternary ammonium-based salts. New Journal of Chemistry. 2009;**33**(1):26-33

[59] Narayan S, Muldoon J, Finn MG, Fokin VV, Kolb HC, Sharpless KB. "On Water": unique reactivity of organic compounds in aqueous suspension. Angewandte Chemie International Edition. 2005;**44**:3275-3279

[60] Sharma SR. Green chemistry, green solvents and alternative techniques in organic synthesis. International Journal of Chemical and Physical Sciences. 2015;**4**:516-520. (Special Issue-NCSC-Jan-2015)

[61] Breslow R. The principles of and reasons for using water as a solvent for green chemistry. In: Part 5. Reactions in Water Published Online. Germany: Wiley-VCH Verlag GmbH & Co. KGaA; 2010

[62] Chanda A, Fokin VV. Oragnic synthesis "on water". Chemical Reviews. 2009;**109**(2):725-748

[63] Narayan S, Fokin V, Sharpless K. Organic reactions in water. In: Lindstrom M editor. Principles, Strategies and Applications. Oxford: Blackwell; 2007. pp. 350-365

[64] Breslow R, Maitra U. On the origin of product selectivity in aqueous diels-alder reactions. Tetrahedron Letters. 1984;**25**:1239-1240

[65] Abbott AP, Capper G, Davies DL, Rasheed RK, Tambyrajah V. Novel solvent properties of choline chloride/urea mixtures. Chemical Communications. 2003;**39**:70-71

[66] Guthrie F. LII. On eutexia. The London, Edinburgh, and Dublin

Philosophical Magazine and Journal of Science. 1884;**17**:462-482

[67] Martins MA, Pinho SP, Coutinho JAP. Insights into the nature of eutectic and deep eutectic mixtures. Journal of Solution Chemistry. 2019;**48**: 962-982

[68] Perna FM, Vitale P, Capriati V. Deep eutectic solvents and their applications as green solvents. Current Opinion in Green and Sustainable Chemistry. 2020;**21**:27-33

[69] Florindo C, Lima F, Ribeiro BD, Marrucho IM. Deep eutectic solvents: overcoming 21st century challenges. Current Opinion in Green and Sustainable Chemistry. 2019;**18**:31-36

[70] Paiva A, Matias AA, Duarte ARC. How do we drive deep eutectic systems towards an industrial reality? Current Opinion in Green and Sustainable Chemistry. 2018;**11**:81-85

[71] Smith EL, Abbott AP, Ryder KS. Deep eutectic solvents (DESs) and their applications. Chemical Reviews. 2014;**114**:11060-11082

[72] Choi YH, van Spronsen J, Dai Y, Verberne M, Hollmann F, Arends IW, et al. Are natural deep eutectic solvents the missing link in understanding cellular metabolism and physiology? Plant Physiology. 2011;**156**:1701-1705

[73] Santos LB, Assis RS, Barreto JA, Bezerra MA, Novaes CG, Lemos VA. Deep eutectic solvents in liquid-phase microextraction: contribution to green chemistry. TrAC Trends in Analytical Chemistry. 2021;**146**:116478

[74] O'Brien RD. Fats and Oils: Formulating and Processing for Applications. 3rd ed. Boca Raton, USA: CRC Press; 2005

[75] Menges N, Şhin E. Metal- and base-free combinatorial reaction for C-acylation of 1,3-diketocompounds in vegetable oil: The effect of natural oil. ACS Sustainable Chemistry & Engineering. 2014;**2**:226-230

[76] Branch JA, Bartlett PN. Electrochemistry in supercritical fluid. Philosophical Transactions of the Royal Society A: Mathematical, Physical and Engineering Sciences. 2015;**373**(2057):20150007

[77] Nalawade SP, Picchioni F, Janssen LPBM. Supercritical carbon dioxide as a green solvent for processing polymer melts: processing aspects and applications. Progress in Polymer Science. 2006;**31**(1):19-43

[78] Dharaskar SA. The green solvents for petroleum and hydrocarbon industries. Research Journal of Chemical Sciences. 2012;**2**(8):80-84

[79] Mallakpour S, Rafiee Z. Fundamental and industrial application. In: Inamuddin MA editor. Green Solvents. Millersville, PA, USA: Springer; 2012. pp. 10-50. Ch. 1

[80] Hussain EA. Green solvents for drug synthesis, green sustainable process for chemical and environmental engineering and science. In: Inamuddin R, Boddula MI, Ahamed AM, Asiri AM, editors. Solvents for the Pharmaceutical Industry. Amsterdam, Netherlands: Elsevier; 2021. pp. 55-86. Ch. 4

[81] Choi YH, Verpoorte R. Green solvents for the extraction of bioactive compounds from natural products using ionic liquids and deep eutectic solvents. Current Opinion in Food Science. 2019;**26**:87-93

[82] Truong CC, Mishra DK, Mishra V. Organic carbonate as a green solvent for

biocatalysis. In: Inamuddin, Boddula R, Ahamed MI, Asiri AM, editors. Green Sustainable Process for Chemical and Environmental Engineering and Science. Amsterdam, Netherlands: Elsevier; 2021. pp. 253-270. Ch. 13

[83] Prasad K, Sharma M. Green solvents for the dissolution and processing of biopolymers. Current Opinion in Green and Sustainable Chemistry. 2019;**18**:72-78

[84] Arrachart G, Couturier J, Dourdain S, Levard C, Pellet-Rostaing S. Recovery of Rare Earth Elements (REEs) Using Ionic Solvents. Multidisciplinary Digital Publishing Institute. 2021;**9**:1202

[85] Díaz-Álvarez AE, Cadierno V. Glycerol: a promising green solvent and reducing agent for metal-catalyzed transfer hydrogenation reactions and nanoparticles formation. Applied Sciences. 2013;**3**(1):55-69

[86] Jiang Y, Li J, Feng ZW, Xu G, Shi X, Ding QJ, et al. Ethylene glycol: a green solvent for visible light-promoted aerobic transition metal-free cascade sulfonation/cyclization reaction. Advanced Synthesis and Catalysis. 2020;**362**(13):2609-2614

[87] Berthod A, Carda-Broch S. A new class of solvents for CCC: the room temperature ionic liquids. Journal of Liquid Chromatography and Related Technologies. 2003;**26**:1493-1508

[88] Marszall MP, Kaliszan R. Application of ionic liquids in liquid chromatography. Critical Reviews in Analytical Chemistry. 2007;**37**:127-140

[89] Merike V, Mihkel K, Mihkel K. Ionic liquids as electrolytes for nonaqueous capillary electrophoresis. Electrophoresis. 2002;**23**:426-430

[90] Yanes EG, Gratz SR, Baldwin MJ, Robison SE, Stalcup AM. Capillary electrophoretic application of 1-alkyl-3-methylimidazolium-based ionic liquids. Analytical Chemistry. 2001;**73**:3838-3844

[91] López-Pastor M, Simonet BM, Lendl B, Valcárcel M. Ionic liquids and CE combination. Electrophoresis. 2008;**29**:94-107

[92] Yue ME, Shi YP. Application of 1-alkyl-3-methylimidazolium-based ionic liquids in separation of bioactive flavonoids by capillary zone electrophoresis. Journal of Separation Science. 2006;**29**:272-276

[93] François y, Varenne A, Juillerat E, Servais AC, Chiap P, Gareil P. Nonaqueous capillary electrophoretic behavior of 2-aryl propionic acids in the presence of an achiral ionic liquid a chemometric approach. Journal of Chromatography. A. 2007;**1138**:268-275

[94] Ventura SP, Fa ES, Quental MV, Mondal D, Freire MG, Coutinho JA. Ionic-liquid mediated extraction and separation processes for bioactive compounds: past, present, and future trends. Chemical Reviews. 2017;**117**:6984. DOI: 10.1021/acs.chemrev.6b00550

[95] Zhao CN, Zhang JJ, Li Y, Meng X, Li HB. Microwave- assisted extraction of phenolic compounds from Melastoma sanguineum fruit: optimization and identification. Molecules. 2018;**23**:2498. DOI: 10.3390/molecules2310249

[96] Chemat F, Vian MA, Cravotto G. Green extraction of natural products: conceptand principles. International Journal of Molecular Sciences. 2012;**13**:8615. DOI: 10.3390/ijms13078615

[97] Ibañez E, Cifuentes A. Green Extraction Techniques: Principles, Advances and Applications. Amsterdam: Elsevier; 2017

[98] Soquetta MB, Terra LM, Bastos CP. Green technologies

for the extraction of bioactive compounds in fruits and vegetables. CyTA – Journal of Food. 2018;**16**:400. DOI: 10.1080/19476337.2017.1411978

[99] Chemat F, Vian MA. Alternative Solvents for Natural Products Extraction, Green Chemistry and Sustainable Technology. Berlin, Heidelberg: Springer Verlag; 2014. DOI: 10.1007/978-3-662-43628-8

[100] Torres-Valenzuela LS, Ballesteros-Gomez A, Rubio S. Green solvents for the extraction of high added-value compounds from agri-food waste. Food Engineering Reviews. 2020;**12**:83. DOI: 10.1007/s12393-019-09206-y

[101] Nkuku CA, LeSuer RJ. Journal of Physical Chemistry. 2007;**11**:13271

[102] Zhao H, Baker GA. Journal of Chemical Technology & Biotechnology. 2013;**88**:3

[103] Available from: https://axial.acs.org/2018/04/02/green-solvents-embrace-future [Last accessed: 25 Sep 2021]

[104] Pace V, Hoyos P, Castoldi P, de Mara D, Alcntara AR. 2-methyltetrahydrofuran (2-MeTHF): a biomass-derived solvent with broad application in organic chemistry. ChemSusChem. 2012;**5**(8):1369-1377

[105] Forero JSB, Muñoz JH, Jones J, da Silva JFM. Propylene carbonate in organic synthesis: exploring its potential as a green solvent. Current Organic Synthesis. 2016;**13**:834-844

[106] Bermejo DV, Mendiola JA, Ibánez E, Reglero G, Fornari T. Pressurized liquid extraction of caffeine and catechins from green tea leaves using ethyl lactate, water and ethyl lactate +water mixtures. Food and Bioproducts Processing. 2015;**96**:106-111

Chapter 2

Green Solvents in Organic Synthesis

Geeta Verma

Abstract

Solvents are substances that are liquid during application and will dissolve other substances, which can be recovered unchanged on the removal of the solvent. The concept of green solvents indicates the target to decrease the environmental impact resulting from the use of solvents in chemical production. Green solvents are solvents of low toxicity, environment-friendly bio solvents, and less hazardous than traditional organic solvents. The solvents which are not harmful to the environment and human beings are called green solvents. Traditional solvents can be replaced with green solvents as a long-term way to reduce and minimize environmental deterioration. Solvents like ionic liquids and deep eutectic mixtures can be used as green solvents and are used as part of the class of green solvents in organic synthesis. The review focuses on the properties, applications, and limitations of these solvents.

Keywords: eutectic mixtures, water, ionic liquids, sustainable, neurotransmitter

1. Introduction

1.1 Green solvents

What are green solvents...

Green solvents, commonly referred to as environmentally friendly solvents, are products of crop processing.

The United Nations established a new development strategy with an emphasis on sustainability in 2015, based on 17 Sustainable development goals. It acknowledges the necessity of "green chemistry" and "green solvent" **Figure 1** for future chemistry that is more environmentally friendly. So-called green, ecological, biodegradable, and sustainable solvents have arisen in this setting [1]. which were created as a less harmful alternative to petrochemical solvents known as biosolvents, which are generated from the processing of agricultural crops.

1.2 Types of green solvent

1.2.1 Ionic liquid

The term "ionic liquid" is very vast, ionic liquids (ILs) are made solely of ions that belong to a class of non-molecular compounds having melting points lower

IntechOpen

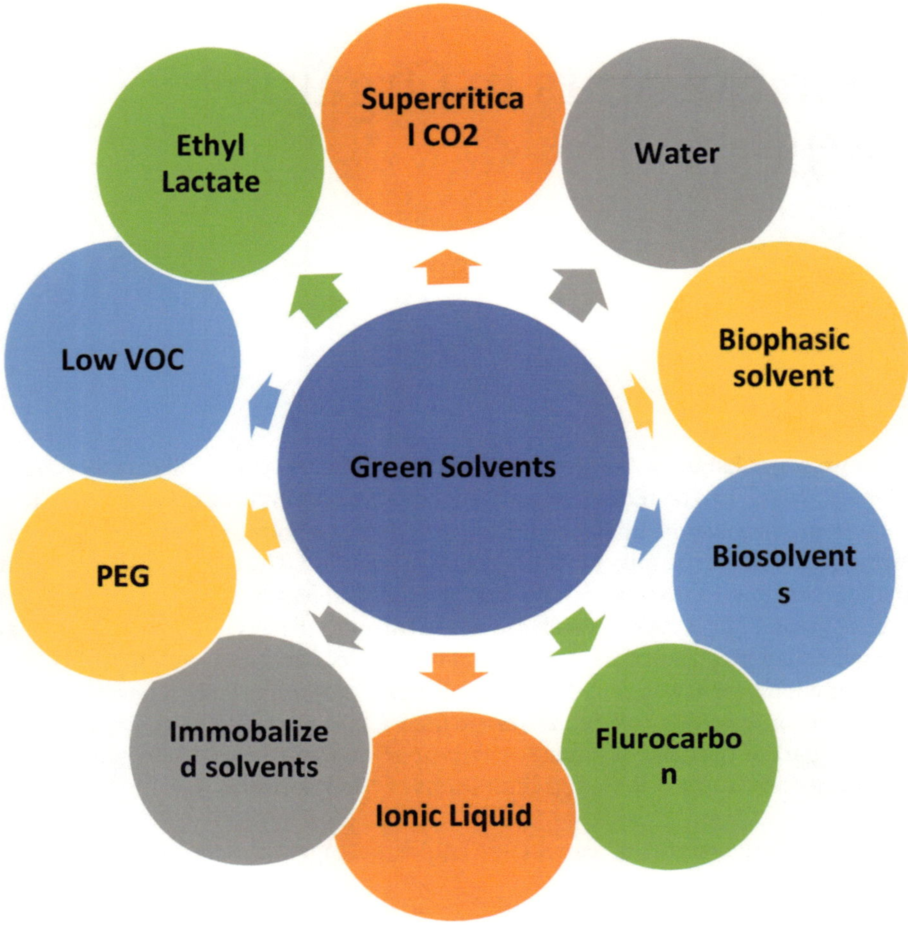

Figure 1.
Types of green solvents.

than 100°C. ILs have several advantages over traditional organic solvents. ILs possess negligible vapor pressure at room temperature as well as high thermal stability and play an important role as ideal solvents in various extraction techniques. The physical and chemical properties of ILs can be varied by simply changing the combination of cations and anions for e.g. viscosity, thermal stability, and solubility in water as well as in other organic solvents.

Over the past two decades, ILs have emerged as a class of promising solvents with unique properties that have several applications in organic synthesis, electrochemistry, catalysis, separation of metals, gas separation, energy storage devices, biomass processing, pharmaceuticals, and tribology. Ionic liquids also recognized by several different names like neoteric solvents, designer solvents, ionic fluids, and molten salts.

In 1914 the first room temperature ionic liquid [$EtNH_3$][NO_3] (m.p. 12°C) was discovered [1].

The behavior of ionic liquid may be an acidic, basic, or organocatalyst and it is determined by the presence of functional group attached to the cation and/or anion.

1.2.2 Ionic liquid for anhydrous reaction condition

It has been reported that ionic liquids [C_4C_1im][HSO_4] and [C_4C_1im][$MeSO_3$] can be successfully used for lignocellulosic biomass treatment even in the presence of significant amount of water, in this way use of ILs eliminates the requirement for anhydrous conditions during pretreatment [2].

The saccharification of cellulose and its successive transformation into significant molecules like hydroxymethylfurfural, levulinic acid and furfural, has been well examined and achieved by the use of acidic ionic liquids [3–9].

1.2.3 Ionic liquid as organocatalyst

Synthesis of sulfur functionalized chiral ionic liquid as an organocatalyst which has been used for the preparation of trans epoxide with high diastereoselectivity and enantioselectivity up to 72% ee, from various aromatic aldehydes with benzyl bromide in aqueous condition.

1.2.4 ILs as support for catalyst/reagents

Importance of ionic liquids as soluble supports for catalyst/reagent immobilization [10] is well studied in various ionic liquid supported synthesis (ILSS) and is applied for a number of organic reactions like Knoevengeal reaction [11], 1,3-cycloadditions [12], oligosaccharide synthesis [13], Suzuki coupling [14], synthesis of thiazolidinones [15], and Grieco's multicomponent synthesis of tetrahydroquinolines [16]. All of the reported reactions are due to controlled solubility and nonvolatile nature of ILs.

The major drawbacks of common ILs, namely high toxicity, non-biodegradability, complex synthesis requiring purification, and high cost of the starting materials [17–32].

1.3 Deep Eutectice Solvents (DESs)

Deep Eutectic Solvents (DESs), are defined as combinations of non ideal mixture of two biodegradable constituents (HBA and HBD) (**Table 1**) with strong hydrogen bonding interactions. They are also known in the literature as Deep Eutectic Ionic Liquids (DEILs), Low Melting Mixtures (LMMs), or Natural Deep Eutectic Solvents (NADES), Low Transition Temperature Mixtures (LTTMs) [33].

A highly non-ideal mixture of two biodegradable constituents (HBA and HBD) with strong hydrogen bonding interactions is categorized as a Deep Eutectic Solvent (DES).

According to this definition, many DESs have been synthesized and has found much application as a solvent in various organic reactions, extraction of dyes, protein, nucleic acids, metals, azeotropic separation, and more [34–38].

The concept "Deep Eutectic Solvent" was first coined by Abbott et al. [39] to describe the formation of a liquid eutectic mixture (melting point 12°C) starting from two solid materials with high melting points: (i) choline chloride (ChCl), (2-hydroxyethyl)trimethylammonium chloride, melting point 133°C, and (ii) urea (melting point 302°C) in a ratio 1:2 (1$ChCl_2$Urea) [40, 41].

A method for determining the polarity and hydrophobicity of DESs was developed by Cichocki et al. [42]. This approach eliminates the necessity for solution preparation, thus in this way does not alter the internal structure of DESs. The presence of

HYDROGEN BONHBA	HYDROGEN BOND DONORs HBD
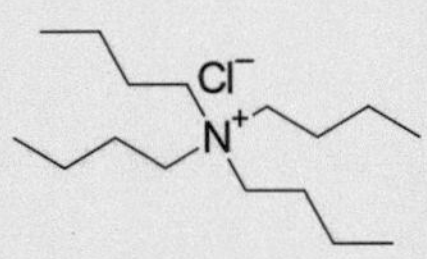 tetra butylammonium chloride	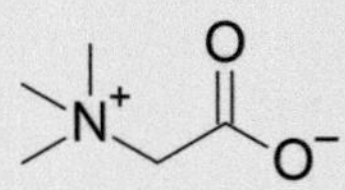 betaine 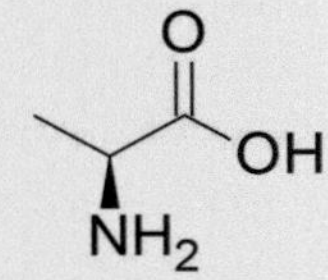 alanine
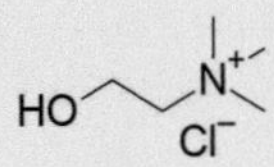 Choline chloride	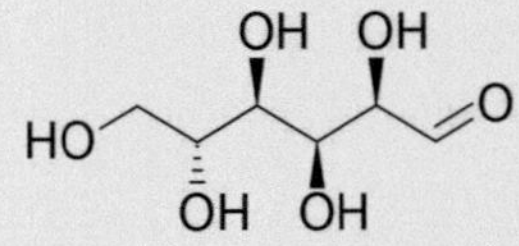 Glucose
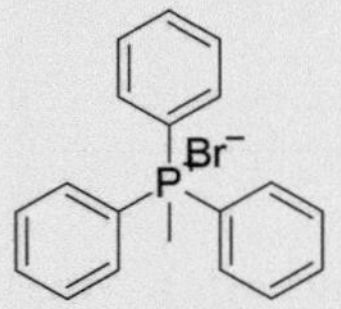 methyltriphenylphosphonium bromide	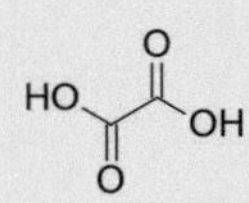 Oxalic acid
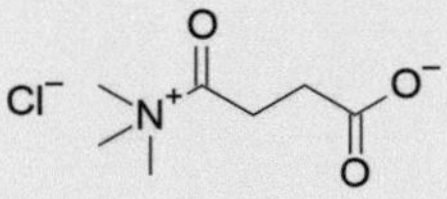 2 acetate N N N trimethylethanamidium chloride	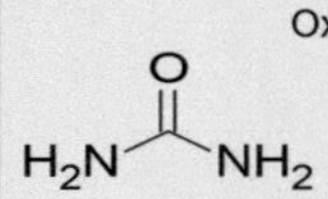 Urea
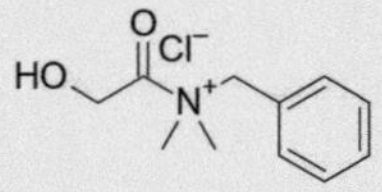 N benzyl 2 hydroxy N N dimethylethanamidium chloride	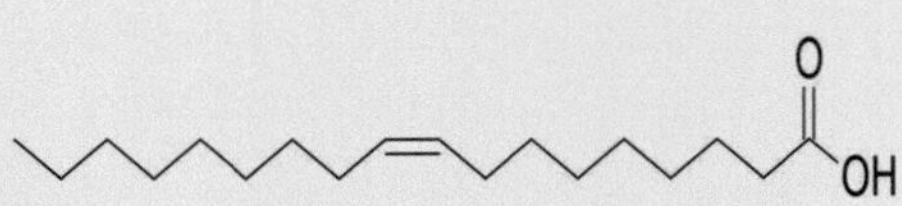 oleic acid
	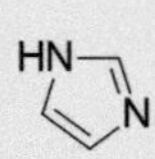 Imidazole 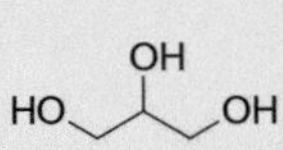 Glycerol

Table 1.
Types of hydrogen bond acceptor (HBA) and hydrogen bond donor (HBD).

an extra component, like as water or another solvent, can cause modifications in the DES molecule's hydrogen bonding mechanism. The contact angle test, which is one way to determine DESs hydrophobicity without having to prepare their solutions.

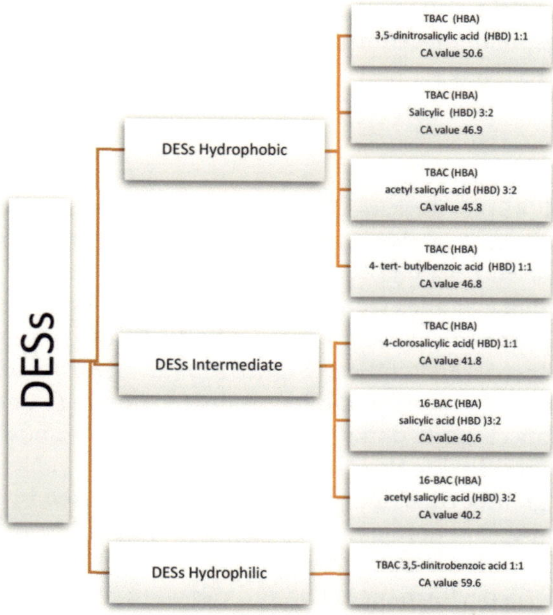

Figure 2.
Classification of DES properties by contact angle (CA) test.

The method does not require the use of any solvents, only an optical goniometer, and a reference surface.

For classifying DESs properties as hydrophobic and hydrophilic, their contact angle values obtained on glass were taken into consideration with respect to highly hydrophobic rapeseed oil and highly hydrophilic water, respectively as a reference substance. DESs molecules in which contact angle value were found to be in between water and oil contact angle value, were classify as having intermediate properties. The contact angle, therefore, indirectly measures the closeness of DES molecules to the surface of the reference substance. The results of contact angle measurements are summarized in **Figure 2**.

1.3.1 Types of Deep Eutectic Solvent (DES)

See **Table 2.**

Type I DES	Quaternary ammonium salt and metal chloride. Imidazolium salts and various metal halides such as $ZnCl_2$, $FeCl_2$, AgCl, $CuCl_2$, $CdCl_2$, LiCl, $SnCl_2$, and $SnCl_4$	[39, 43]
Type II DES	Quaternary ammonium salt and hydrate of metal chloride hydrate	[34]
Type III DES	Quaternary ammonium salt as HBA and HBD. Mainly composed of choline chloride and HBDs (carboxylic acids, alcohols, amides, and carbohydrates, etc.)	[44]
Type IV DES	Metal chloride (particularly transition metal chloride) and HBD	[37, 45]
Type V DES	New class mixture of non-ionic molecular HBA and HBD	[37]

Table 2.
Types of deep eutectic solvent (DES).

DESs were presented using the general formula: $R^{+}A^{-}xB$, where R^{+} is ammonium, sulfonium, and phosphonium cation core. A and B are Lewis base with halide anion and Lewis acid, respectively [37].

1.3.2 DES as catalyst

Azizi and Batebi [46] investigated ChCl-$SnCl_2$ (1:2 molar ratio) as Lewis acid type DES as catalyst for chemoselective ring opening of epoxides with aromatic amines, thiols, alcohols, azide and cyanide.

Synthesis of primary amides from aldehydes and nitriles by Patil et al. [47] was carried out by the used of Lewis -acid type DES ChCl-$ZnCl_2$. Its performance both as a catalyst and solvent were studied (**Figure 3**). Various aromatic, aliphatic and conjugated substrates were used and high product yields were obtained (89–94%). Effect of substituents electron donating groups at *ortho-para* positions on DES catalyzed reactions resulted in excellent yields while *ortho-* substituted groups' reactions required greater time due to steric hindrance caused by substituted groups. Another important application of reusability of DES was examined on the synthesis of benzamide from benzaldehyde. The maximum three recycling has been reported for DES performance. The authors work resulted that the use of DES in the reactions given green and atom-efficient synthesis by reducing waste and toxic material.

Tran et al. [48] used ChCl-$ZnCl_2$ as catalyst and green solvent in Friedel-Crafts acylation reactions. They reported high regio- and chemoselectivity in the reactions using acid anhydrides and ChCl-$ZnCl_2$ (1:3 molar ratio) as catalyst under microwave irradiation (**Figure 4**).

Advantages ChCl-$ZnCl_2$ as catalyst and green solvent in Friedel-Crafts acylation reaction:

- Ketone products having a majority of *p*-isomer
- Prevention of both use of moisture sensitive Lewis acids and volatile organic solvents.
- High yields within short reaction times.

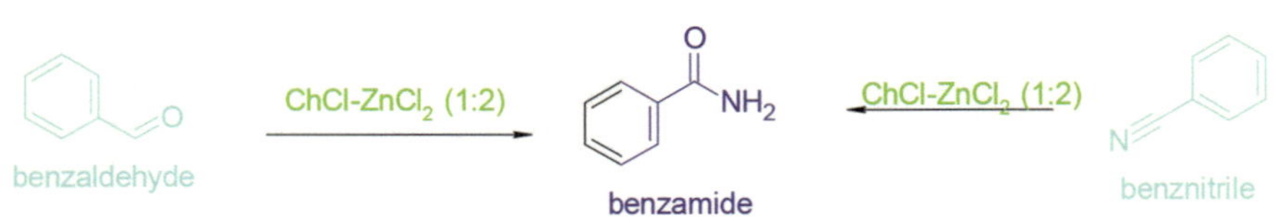

Figure 3.
Synthesis of amides from aldehydes and nitriles catalyzed by DES ChCl-$ZnCl_2$ (1:2).

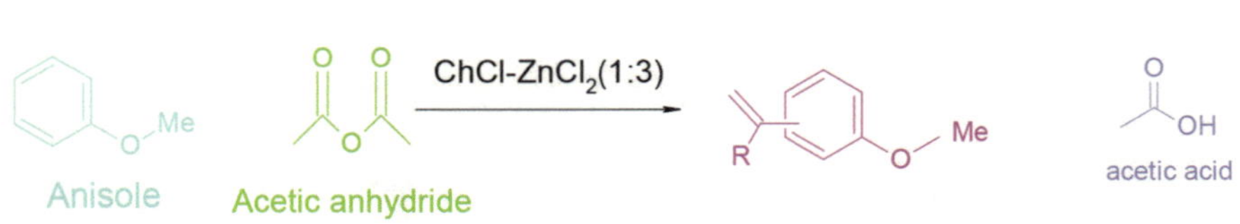

Figure 4.
Friedel craft acylation reaction by DES ChCl-$ZnCl_2$ (1:3).

- Three new ketone products to be synthesized under the catalyzation of DES. ChCl-$ZnCl_2$ (1:3)
- DES could be reused up to 5 times conserving its catalytic activity.

Aromatic ketones are valuable and important precursors for the synthesis of agrochemicals, pharmaceuticals, and also for fragrance as well as dyes etc. The synthetic route for these molecules usually involved Friedel-Crafts acylation using Lewis acids as catalyst [49].

Polyethylene terephthalate (PET) is, the most common thermoplastic polymer resin of the polyester family, extensively used in many fields like in fibres for clothing, containers for liquids and foods, textile industry such as disposable soft drink bottles, packaging and also films and tapes. PET consumption in the world has been reported to surpass 13 million tonnes [40]. Therefore, the need for recycling of PET polymer gains much attention and critical importance for the sake of environment [50].

The glycolysis of PET using DESs as catalyst is firs example reported by Wang et al. [51]. PET degradation was carried out by catalytic reaction using urea-metal salt mixtures' DESs.

H-bonds were between molecules shown in **Table 3**.

In summary, the Wang et al. [51] proposed that DESs mimicked having more catalytic active sites than ionic liquids and metal salts because they could create more H-bonds between various molecules. The reaction's mechanism is emphasized that the H-bonds allow the hydroxyl group's O−H bond in ethylene glycol to lengthen, which increases the oxygen atoms' electronegativity and thus facilitates the loss of hydrogen atoms from oxygen atom and hence resultant greater nucleophilicity. All this factor made it easier for oxygen to attack PET's ester group's carbon. Furthermore, DES and substrate contact was increased by the coordination bonds that existed between PET's oxygen atom and Zn^{+2}. The overall effect of all these factors led to an enhanced PET degradation rate by DESs (**Figure 5**).

H-bonds were between the following molecules

H atom of OH group in ethylene 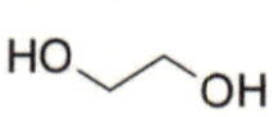 **glycol**	**O atom of CO group in urea** 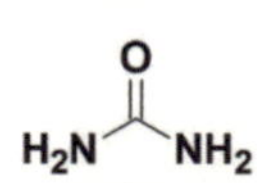
H atom of amidogen group in urea 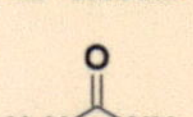	O atom of ester group of PET 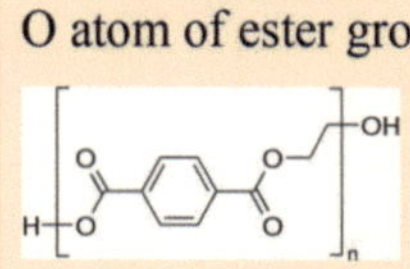
H atom of OH group in ethylene 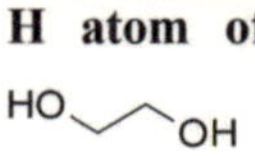 **glycol**	O atom of ester group of PET 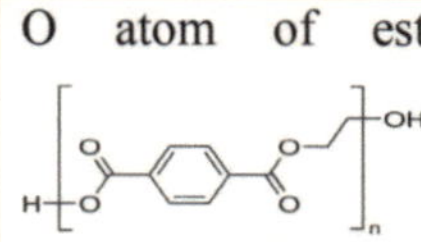
H atom of amidogen group in urea 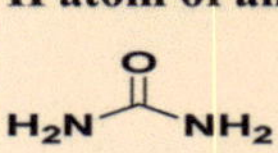	O atom of OH group in ethylene glycol 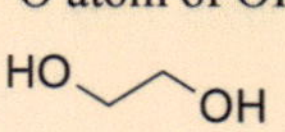

Table 3.
H-bonding between different DES molecules.

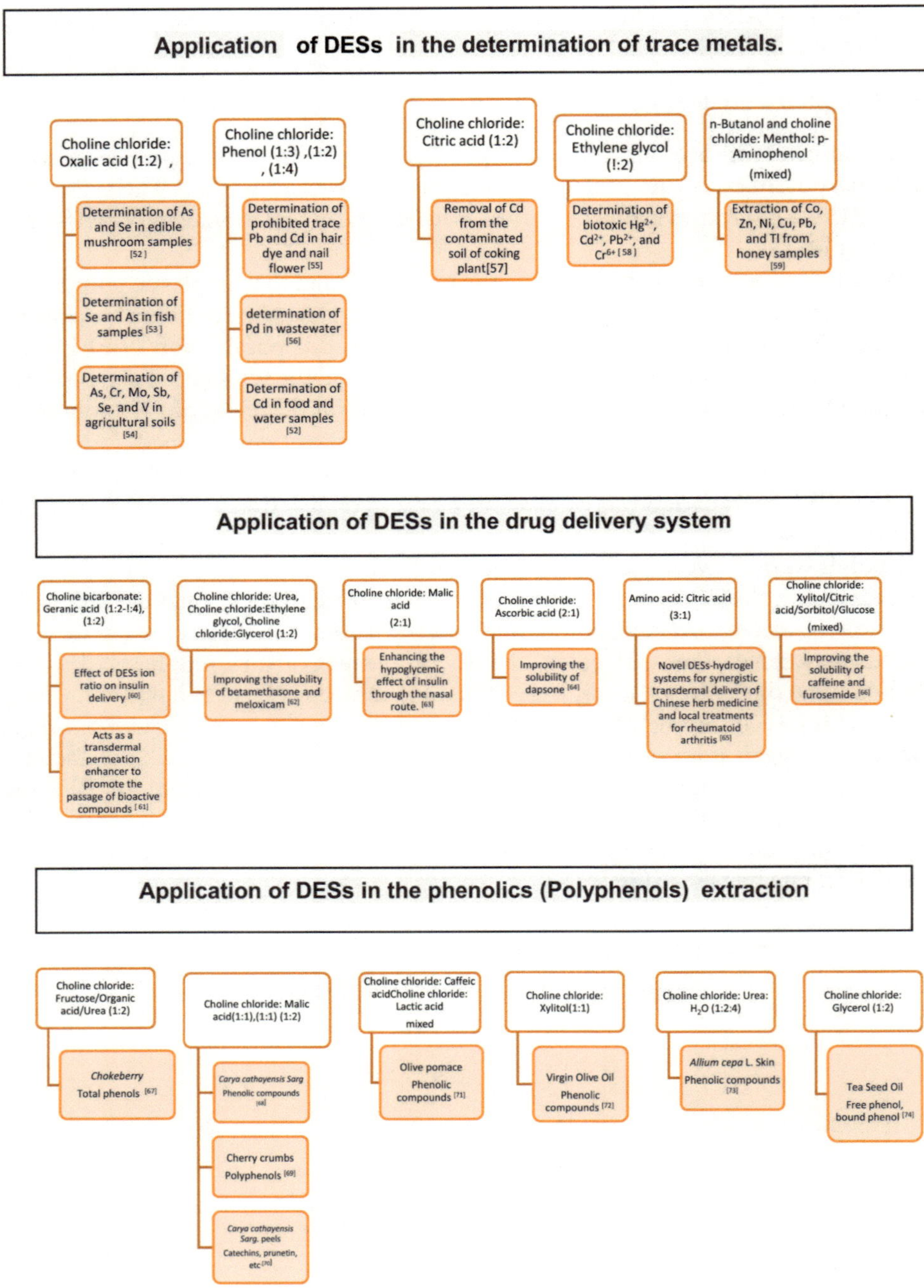

Figure 5.
Application of DES [52–74].

1.3.3 Deep Eutectic Solvents: biomedical applications

Neurotransmitters (NTs) are brain's chemical messengers that are involved in a wide range of neurological processes, including mood, appetite, learning, sleep, and other cognitive processes. Excitatory NTs include epinephrine, norepinephrine, glutamate, serotonin, and acetylcholine; inhibitory NTs include γ-aminobutyric acid, and dopamine (DA) is thought to have both excitatory and inhibitory properties [75]. Severe neuropsychiatric disorders including Parkinson's, schizophrenia, Alzheimer's, and epilepsy can be brought on by an imbalance in the concentration of NT [76].

The application of DESs as green solvents in detecting and extracting NTs is described in detail for the first time [77]. Various methods, such as electrochemical [78], positron-emission tomography [79], optical [80], and microdialysis [81], are available for sensing NTs. Various nanocomposites have been utilized for sensing NTs [82]. Because of its vital characteristics, which include cost-effectiveness, biodegradability, and customizable physiochemical qualities that can control and modify the formation of nanocomposites, DESs can be used to synthesize these nanocomposites [83]. To create nanocomposites, DESs can be used as a precursor, reactant, solvent, and shape-controlling agent.

1.4 Conclusion

With the increasing awareness towards environmental issues, researchers are taking great interest and effort to replace toxic and harmful constituents with less harmful constituents. In this direction Ionic liquids (ILs) and Deep eutectic liquids (DES) have found many successful important synthetic applications in many different fields of research. DESs come upon many applications in various areas such as metal processing, trace metal extraction, drug delivery, extraction of polyphenols and flavonoids, organic synthesis, biotransformations and biomedical application as well. DES was taken as a solvent or co-solvent in various reactions, as it has properties of high solubility or miscibility with the substituents. Therefore, DESs as an acid catalyst can readily be adopted as alternative catalysts to conventional ones.

One of the promising features of DES in various organic synthesis as a catalyst is reusability or recycling nature. Flavonoids and phenolic components posses wide range of pharmacological and medicinal values, so the extraction of flavonoids as well as polyphenols have always been a research hotspot and use of DES can replace toxic organic solvents. Due to its biodegradability, as most of the components that make up DESs are natural products, they can be easily converted into different kinds of organisms in nature. DES can act as both as solvent and as well as catalyst, this kind of dual functionality of biodegradable greener solvent will pave a pathway for organic synthesis in future.

Further research in quest of DES for detecting and extracting NTs is needed to explore the full potential of DESs that could have significant implications for the diagnosis and treatment of neurological disorders and to optimize their use for practical applications. This book chapter provides a brief aspect of some of the most recent applications of ILS and DESs in various fields.

Author details

Geeta Verma
Department of Chemistry, Chandra Shekhar Azad Govt Post Graduate Lead College, Sehore, MP, India

*Address all correspondence to: drgitaverma@yahoo.co.in

References

[1] Walden. Ueber die Molekulargrösse und elektrische Leitfähigkeit einiger geschmolzenen Salze. Bulletin de l'Académie Impériale des Sciences de St.-Pétersbourg. 1914;**8**(6):405-422

[2] Brandt A, Ray MJ, T. Q. To, Leak DJ, Murphy RJ, Welton T. Ionic liquid pretreatment of lignocellulosic biomass with ionic liquid-water mixtures. Green Chemistry. 2011;**13**(9):2489-2499

[3] Amarasekara AS, Owereh OS. Hydrolysis and decomposition of cellulose in bronØsted acidic ionic liquids under mild conditions. Industrial & Engineering Chemistry Research. 2009;**48**(22):10152-10155

[4] Li C, Zhao ZK. Efficient acid-catalyzed hydrolysis of cellulose in ionic liquid. Advanced Synthesis and Catalysis. 2007;**349**(11-12):1847-1850

[5] Zakrzewska ME, Lukasik EB, Lukasik RB. Ionic liquid-mediated formation of 5-hydroxymethylfurfural—a promising biomass-derived building block. Chemical Reviews. 2011;**111**:397-417

[6] Ståhlberg T, Fu W, Woodley JM, Riisager A. Synthesis of 5-(hydroxymethyl)furfural in ionic liquids: Paving the way to renewable chemicals. ChemSusChem. 2011;**4**(4):451-458

[7] Ruiz JCS, Campelo JM, Francavilla M, et al. Efficient microwave-assisted production of furfural from C_5 sugars in aqueous media catalysed by Brönsted acidic ionic liquids. Catalysis Science and Technology. 2012;**2**:1828-1832

[8] Sun Z, Cheng M, Li H, et al. One-pot depolymerization of cellulose into glucose and levulinic acid by heteropolyacid ionic liquid catalysis. RSC Advances. 2012;**2**(24):9058-9065

[9] Ren H, Zhou Y, Liu L. Selective conversion of cellulose to levulinic acid via microwave-assisted synthesis in ionic liquids. Bioresource Technology. 2013;**129**:616-619

[10] Chan TH. Ionic-liquid-supported synthesis: A novel liquid-phase strategy for organic synthesis ionic-liquid-supported synthesis: A novel liquid-phase strategy for organic synthesis. Accounts of Chemical Research. 2006;**39**(12):897-908

[11] Fraga-Dubreuil J, Bazureau JP. Grafted ionic liquid-phase-supported synthesis of small organic molecules. Tetrahedron Letters. 2006;**42**(35):6097-6100

[12] Dubreuil JF, Bazureau JP. Rate accelerations of 1,3-dipolar cycloaddition reactions in ionic liquids. Tetrahedron Letters. 2000;**41**(38):7351-7355

[13] Miao W, Chan TH. Exploration of ionic liquids as soluble supports for organic synthesis. Demonstration with a Suzuki coupling reaction. Organic Letters. 2003;**5**(26):5003-5005

[14] Fraga-Dubreuil J, Bazureau JP. Efficient combination of task-specific ionic liquid and microwave dielectric heating applied to one-pot three component synthesis of a small library of 4-thiazolidinones. Tetrahedron. 2003;**59**(32):6121-6130

[15] Huang J-Y, Lei M, Wang Y-G. A novel and efficient ionic liquid supported synthesis of oligosaccharides. Tetrahedron Letters. 2006;**47**(18):3047-3050

[16] Hassine F, Gmouh S, Pucheault M, Vaultier M. Task specific onium salts and ionic liquids as soluble supports in Grieco's multicomponent synthesis of tetrahydroquinolines. Monatshefte fur Chemie. 2007;**138**(11):1167-1174

[17] Ranke J, Stolte S, Stormann R, Arning J, Jastorff B. Design of sustainable chemical products—The example of ionic liquids. Chemical Reviews. 2007;**107**:2183-2206. DOI: 10.1021/cr050942s

[18] Hough-Troutman WL, Smiglak M, Griffin S, Reichert WM, Mirska I, Jodynis-Liebert J, et al. Ionic liquids with dual biological function: Sweet and anti-microbial, hydrophobic quaternary ammonium-based salts. New Journal of Chemistry. 2009;**33**:26-33. DOI: 10.1039/B813213P

[19] Demberelnyamba D, Kim KS, Choi SJ, Park SY, Lee H, Kim CJ, et al. Synthesis and antimicrobial properties of imidazolium and pyrrolidinonium salts. Bioorganic & Medicinal Chemistry. 2004;**12**:853-857. DOI: 10.1016/j.bmc.2004.01.003

[20] Carson L, Chau PKW, Earle MJ, Gilea MA, Gilmore BF, Gorman SP, et al. Antibiofilm activities of 1-alkyl-3-methylimidazolium chloride ionic liquids. Green Chemistry. 2009;**11**:492-497. DOI: 10.1039/b821842k

[21] Rogers R, Daly DT, Swatloski RP, Hough WL, Davis JH, Smiglak M, et al. Multi-functional ionic liquid compositions. 8,232,265. U.S. Patent. 2006

[22] Kumar V, Malhotra SV. Ionic liquid applications: Pharmaceuticals, therapeutics, and biotechnology. Journal of the American Chemical Society. 2010;**132**:1-12. DOI: 10.1021/ja1098947

[23] Yu M, Li SM, Li XY, Zhang BJ, Wang JJ. Acute effects of 1-octyl-3-methylimidazolium bromide ionic liquid on the antioxidant enzyme system of mouse liver. Ecotoxicology and Environmental Safety. 2008;**71**:903-908. DOI: 10.1016/j.ecoenv.2008.02.022

[24] Fraser KJ, Izgorodina E, Forsyth M, Scott J, Macfarlane DR. Liquids intermediate between "molecular" and "ionic" liquids: Liquid ion pairs? Chemical Communication. 2007:3817-3819. DOI: 10.1039/b710014k

[25] Stolte S, Arning J, Bottin-Weber U, Matzke M, Stock F, Thiele K, et al. Anion effect on cytotoxicity of ionic liquids. Green Chemistry. 2006;**8**:621-629. DOI: 10.1039/b602161a

[26] Fujita K, Forsyth M, MacFarlane DR, Reid RW, Elliott GD. Unexpected improvement in stability and utility of cytochrome c by solution in biocompatible ionic liquids. Biotechnology and Bioengineering. 2006;**94**:1209-1213. DOI: 10.1002/bit.20928

[27] Júlio A, Antunes C, Mineiro R, Raposo M, Caparica R, Araújo MEM, et al. Influence of two choline-based ionic liquids on the solubility of caffeine. Biomedical and Biopharm Research. 2018;**15**:96-102. DOI: 10.19277/bbr.15.1.178

[28] Júlio A, Caparica R, Costa Lima SA, Fernandes AS, Rosado C, Prazeres DMF, et al. Ionic liquid-polymer nanoparticle hybrid systems as new tools to deliver poorly soluble drugs. Nanomaterials. 2019;**9**:1148. DOI: 10.3390/nano9081148

[29] Rogers R, Daly DT, Gurau G, Macfarlane D, Turanjanin J, Dean P, et al. Dual functioning ionic liquids and salts thereof. 9,278,134. U.S. Patent. 2009

[30] Bica K, Rijksen C, Nieuwenhuyzen M, Rogers RD. In search of pure liquid salt forms of aspirin: Ionic liquid approaches with acetylsalicylic acid and salicylic acid. Physical Chemistry. 2010;**12**:2011-2017. DOI: 10.1039/b923855g

[31] Hoffer M, Nutley NJ. Gentisic acid compounds of US 2541651. U.S. Patent. 1951

[32] Cornellas A, Perez L, Comelles F, Ribosa I, Manresa A, Garcia MT. Self-aggregation and antimicrobial activity of imidazolium and pyridinium based ionic liquids in aqueous solution. Journal of Colloid Interface Science. 2011;**355**:164-171. DOI: 10.1016/j.jcis.2010.11.063

[33] Francisco M, van den Bruinhorst A, Kroon MC. Low-transition-temperature mixtures (LTTMs): A new generation of designer solvents. Angewandte Chemical International Edition. 2013;**52**:3074-3075

[34] Abbott AP, Capper G, Davies DL, Rasheed RK, Tambyrajah V. Novel solvent properties of choline chloride/urea mixtures. Chemical Communications. 2003;**1**:70-71

[35] Carriazo D, Serrano MC, Gutierrez MC, Ferrer ML, Monte F. Deep-eutectic solvents playing multiple roles in the synthesis of polymers and related materials. Chemical Society Reviews. 2012;**41**:4996-5014

[36] Pena-Pereira F, Namiesnik J. Ionic liquids and deep eutectic mixtures: Sustainable solvents for extraction processes. ChemSusChem. 2015;7:1-18

[37] Smith EL, Abbott AP, Ryder KS. Deep eutectic solvents (DESs) and their applications. Chemical Reviews. 2014;**114**:11060-11082

[38] García G, Aparicio S, Ullah R, Atilhan M. Deep eutectic solvents: Physicochemical properties and gas separation applications. Energy & Fuels. 2015;**29**:2616-2644

[39] Abbott AP, Capper G, Davies DL, Rasheed RK. Ionic liquid analogues formed from hydrated metal salts. Chemistry - A European Journal. 2004;**10**:3769

[40] George N, Kurian T. Recent developments in the chemical recycling of postconsumer poly (ethylene terephthalate) waste. Industrial and Engineering Chemistry Research. 2014;**53**:14185-14198. DOI: 10.1021/ie501995

[41] Abbott AP, Capper G, Davies DL, Rasheed RK. Ionic liquids based upon metal halide/substituted quaternary ammonium salt mixtures. Inorganic Chemistry. 2004;**43**:3447

[42] Cichocki L, Warmińska D, Łuczak J, Przyjazny A, Boczkaj G. New simple and robust method for determination of polarity of deep eutectic solvents (DESs) by means of contact angle measurement Preprints.org. 10.20944/preprints202205.0139.v1

[43] Xu WG, Lu XM, Zhang QG, Gui JS, Yang JZ. Study on thermodynamic properties of ionic liquids BMIMGaCl2. Chinese Journal of Chemistry. 2006;**24**:331

[44] Gambino M, Bros JP. Heat capacity of urea and of a group of eutectic mixtures based on urea between 30 and 140 °C. Thermochimica Acta. 1988;**127**:223-236

[45] Abranches DO, Martins MA, Silva LP, Schaeffer N, Pinho SP, Coutinho JAP. Phenolic hydrogen bond donors in the formation of non-ionic deep eutectic solvents: The quest for type V DES. Chemical Communications. 2019;**55**:10253-10256

[46] Azizi N, Batebi E. Highly efficient deep eutectic solvent catalyzed ring opening of epoxides. Catalysis Science & Technology. 2012;**2**:2445-2448

[47] Patil UB, Singh AS, Nagarkar JM. Choline chloride based eutectic solvent: An efficient and reusable solvent system for the synthesis of primary amides from aldehydes and from nitriles. RSC Advances. 2014;**4**:1102-1106

[48] Tran PH, Nguyen HT, Hansen PE, Le TN. An efficient and green method for regio-and chemo-selective Friedel–Crafts acylations using a deep eutectic solvent ([CholineCl][ZnCl 2] 3). RSC Advances. 2016;**6**:37031-37038

[49] Chakraborti AK, Gulhane R. Perchloric acid adsorbed on silica gel as a new, highly efficient, and versatile catalyst for acetylation of phenols, thiols, alcohols, and amines. Chemical Communications. 2003;**15**:1896-1897. DOI: 10.1039/B304178F

[50] Musale RM, Shukla SR. Deep eutectic solvent as effective catalyst for aminolysis of polyethylene terephthalate (PET) waste. International Journal of Plastics Technology. 2016;**20**:106-120. DOI: 10.1007/s12588-016-9134-7

[51] Wang A, Xing P, Zheng X, Cao H, Yang G, Zheng X. Deep eutectic solvent catalyzed Friedel–Crafts alkylation of electron-rich arenes with aldehydes. RSC Advances. 2015;**5**:59022-59026. DOI: 10.1039/C5RA08950F

[52] Zounr RA, Tuzen M, Deligonul N, Khuhawar MY. A highly selective and sensitive ultrasonic assisted dispersive liquid phase microextraction based on deep eutectic solvent for determination of cadmium in food and water samples prior to electrothermal atomic absorption spectrometry. Food Chemistry. 2018;**253**:277-283

[53] Panhwar AH, Tuzen M, Kazi TG. Choline chloride–oxalic acid as a deep eutectic solvent–based innovative digestion method for the determination of selenium and arsenic in fish samples. Journal of AOAC International. 2018;**101**:1183-1189

[54] Baghaei PA, Mogaddam MR, Farajzadeh MA, Mohebbi A, Sorouraddin SM. Application of deep eutectic solvent functionalized cobalt ferrite nanoparticles in dispersive micro solid phase extraction of some heavy metals from aqueous samples prior to ICP-OES. Journal of Food Composition and Analysis. 2023;**117**:105125

[55] Yahya M, Kesekler S, Durukan İ, Arpa Ç. Determination of prohibited lead and cadmium traces in hair dyes and henna samples using ultrasound assisted-deep eutectic solvent-based liquid phase microextraction followed by microsampling-flame atomic absorption spectrometry. Analytical Methods. 2021;**13**:1058-1068

[56] Koçoğlu ES, Yılmaz Ö, Bakırdere EG, Bakırdere S. Quantification of palladium in wastewater samples by matrix-matching calibration strategy assisted deep eutectic solvent based microextraction. Environmental Monitoring and Assessment. 2021;**193**:344-360

[57] Qian J, Li Y-H, Su F, Wu J-G, Sun J-R, Huang T-C. Citric acid-based deep eutectic solvent (CA-DES) as a new soil detergent for the removal of cadmium from coking sites. Environmental Science and Pollution Research International. 2022;**30**:2118-2127

[58] da Silva WW, Ghica ME, Brett CMA. Biotoxic trace metal ion detection by enzymatic inhibition of a glucose biosensor based on a poly (brilliant green)-deep eutectic solvent/carbon

nanotube modified electrode. Talanta. 2020;**208**:120-127

[59] Farisi P, Mogaddam MR, Farajzadeh MA, Nemati M. Development of salt-induced homogenous liquid-liquid extraction based on ternary deep eutectic solvent coupled with dispersive liquid-liquid microextraction for the determination of heavy metals in honey. Journal of Food Composition and Analysis. 2023;**117**:105107

[60] Tanner EEL, Ibsen KN, Mitragotri S. Transdermal insulin delivery using choline-based ionic liquids (CAGE). Journal of Controlled Release. 2018;**286**:137-144

[61] Boscariol R, Caetano A, Silva EC, Oliveira TJ, Rosa-Castro RM, Vila MMDC, et al. Performance of choline Geranate deep eutectic solvent as transdermal permeation enhancer: An In vitro skin histological study. Pharmaceutics. 2021;**13**:540

[62] Salva G, Masumeh M, Hemayat K. Solubility enhancement of betamethasone, meloxicam and piroxicam by use of choline-based deep eutectic solvents. Pharmaceutical Science. 2021;**27**:86-101

[63] Li Y, Wu X, Zhu Q, Chen Z, Lu Y, Qi J, et al. Improving the hypoglycemic effect of insulin via the nasal administration of deep eutectic solvents. International Journal of Pharmaceutics. 2019;**56**:118-584

[64] Trombino S, Siciliano C, Procopio D, Curcio F, Laganà AS, Di Gioia ML, et al. Deep eutectic solvents for improving the solubilization and delivery of Dapsone. Pharmaceutics. 2022;**14**:333

[65] Xiao S, Wang L, Han W, Gu L, Cui X, Wang C. Novel deep eutectic solvent-hydrogel systems for synergistic transdermal delivery of Chinese herb medicine and local treatments for rheumatoid arthritis. Pharmaceutical Research. 2022;**39**:2431-2446

[66] Lomba L, Polo A, Alejandre J, Martínez N, Giner B. Solubility enhancement of caffeine and furosemide using deep eutectic solvents formed by choline chloride and xylitol, citric acid, sorbitol or glucose. Journal of Drug Delivery Science and Technology. 2023;**79**:104010

[67] Islamčević Razboršek M, Ivanović M, Krajnc P, Kolar M. Choline chloride based natural deep eutectic solvents as extraction media for extracting phenolic compounds from chokeberry (*Aronia melanocarpa*). Molecules. 2020;**25**:1619

[68] Fu X, Wang D, Belwal T, Xu Y, Li L, Luo Z. Sonication-synergistic natural deep eutectic solvent as a green and efficient approach for extraction of phenolic compounds from peels of *Carya cathayensis* Sarg. Food Chemistry. 2021;**355**:375-395

[69] Popovic BM, Micic N, Potkonjak A, Blagojevic B, Pavlovic K, Milanov D, et al. Novel extraction of polyphenols from sour cherry pomace using natural deep eutectic solvents-ultrafast microwave-assisted NADES preparation and extraction. Food Chemistry. 2022;**366**:958-989

[70] Fu X, Belwal T, He Y, Xu Y, Li L, Luo Z. UPLC-triple-TOF/MS characterization of phenolic constituents and the influence of natural deep eutectic solvents on extraction of Carya cathayensis Sarg. peels: Composition, extraction mechanism and in vitro biological activities. Food Chemistry. 2022;**370**:131-154

[71] Chanioti S, Katsouli M, Tzia C. Novel processes for the extraction of phenolic compounds from olive pomace and their protection by encapsulation. Molecules. 2021;**26**:1781

[72] Rodríguez-Juan E, Rodríguez-Romero C, Fernández-Bolaños J, Florido MC, Garcia-Borrego A. Phenolic compounds from virgin olive oil obtained by natural deep eutectic solvent (NADES): Effect of the extraction and recovery conditions. Journal of Food Science and Technology. 2021;**58**:552-561

[73] Bhushan PC, Girirajsinh CJ. Microwave-assisted deep eutectic solvent extractionof phenolic antioxidants from onion (*Allium cepa* L.) peel: A Box-Behnken design approach for optimization. Journal of Food Science and Technology. 2019;**56**:238-265

[74] Wang X, Jia W, Lai G, Wang L, Contreras MDM, Yang D. Extraction for profiling free and bound phenolic compounds in tea seed oil by deep eutectic solvents. Journal of Food Science. 2020;**85**:284-349

[75] Perry E, Ashton H, Young A. Neurochemistry of Consciousness: Neurotransmitters in Mind. Advances in Consciousness Research. Amsterdam, The Netherlands: John Benjamins Publishing Company; 2002

[76] Wang P, Li Y, Huang X, Wang L. Fabrication of layer-by-layer modified multilayer films containing choline and gold nanoparticles and its sensing application for electrochemical determination of dopamine and uric acid. Talanta. 2007;**73**(3):431-437. DOI: 10.1016/j.talanta.2007.04.022

[77] Kaur H, Siwal SS, Kumar V, Thakur VK. Deep eutectic solvents toward the detection and extraction of neurotransmitters: An emerging paradigm for biomedical applications. Industrial and Engineering Chemistry Research. 2023;**62**(45):18906-18917

[78] Tavakolian-Ardakani Z, Hosu O, Cristea C, Mazloum-Ardakani M, Marrazza G. Latest trends in electrochemical sensors for neurotransmitters: A review. Sensors. 2019;**19**(9):2037. DOI: 10.3390/s19092037

[79] Finnema SJ, Scheinin M, Shahid M, Lehto J, Borroni E, Bang-Andersen B, et al. Application of cross-species PET imaging to assess neurotransmitter release in brain. Psychopharmacology. 2015;**232**(21):4129-4157. DOI: 10.1007/s00213-015-3938-6

[80] Soleymani J. Advanced materials for optical sensing and biosensing of neurotransmitters. TrAC Trends in Analytical Chemistry. 2015;**72**:27-44. DOI: 10.1016/j.trac.2015.03.017

[81] Zestos AG, Kennedy RT. Microdialysis coupled with LC-MS/MS for in vivo neurochemical monitoring. AAPS Journal. 2017;**19**(5):1284-1293. DOI: 10.1208/s12248-017-0114-4

[82] Yang J, Cheng S, Qin S, Huang L, Xu Y, Wang Y. CeO_2–Co_3O_4 nanocomposite with oxidase-like activity for colorimetric detection of ascorbic acid. RSC Advances. 2023;**13**(15):9918-9923. DOI: 10.1039/D3RA01074K

[83] Munyemana JC, Chen J, Wei X, Ali MC, Han Y, Qiu H. Deep eutectic solvent-assisted facile synthesis of copper hydroxide nitrate nanosheets as recyclable enzyme-mimicking colorimetric sensor of biothiols. Analytical and Bioanalytical Chemistry. 2020;**412**(19):4629-4638. DOI: 10.1007/s00216-020-02712-7

Chapter 3

Deep Eutectic Solvents as a New Frontier in Drilling Fluid Design: Opportunities and Challenges

Muhammad Hammad Rasool and Maqsood Ahmad

Abstract

The chapter covers the topic of deep eutectic solvents (DESs) and their potential as drilling fluid additives, focusing on their use as mud rheology modifiers, shale swelling inhibitors, and hydrate inhibitors during drilling. The properties and characteristics of DES, as well as their advantages and disadvantages, are discussed in detail. The environmental benefits of DES-based drilling fluids, compatibility with other drilling fluid additives, and potential future applications are also examined. Additionally, the challenges and limitations of using DESs as a drilling fluid additive and the safety concerns associated with their use are highlighted. Overall, this chapter provides a comprehensive overview of the use of DES in drilling fluids and its potential to improve the efficiency and safety of drilling operations.

Keywords: deep eutectic solvents, drilling fluids, mud rheology, shale inhibitor, hydrate inhibitor

1. Introduction

1.1 Overview of drilling fluids and their importance in drilling operations

Drilling fluids, also known as drilling muds, are an essential component in drilling operations. They are used to lubricate and cool the drill bit, suspend cuttings and debris, and control pressure in the wellbore. Without drilling fluids, drilling operations would be impossible to carry out, as they play a critical role in ensuring the safety and efficiency of the drilling process. Drilling fluids are composed of various components, including a base fluid, which can be water, oil, or synthetic oil, and additives such as clays, polymers, and weighting agents. The type of drilling fluid used depends on the type of well being drilled, the formation being penetrated, and the conditions of the drilling operation. The composition of the drilling fluid is carefully selected to meet the specific requirements of each drilling operation [1].

One of the most important functions of drilling fluids is to provide stability to the wellbore. As the drill bit penetrates the formation, the drilling fluids help to keep the wellbore stable, preventing the formation from collapsing in on itself. This is particularly important when drilling through unstable or unconsolidated formations, where the pressure of the drilling fluid helps to support the formation and prevent it from collapsing. Another crucial function of drilling fluids is to cool and lubricate the drill bit. Drilling generates a significant amount of heat, which can cause damage to the bit and decrease drilling efficiency. Drilling fluids help to cool the bit by circulating around it and carrying away the heat generated by the drilling process. Additionally, they also provide lubrication, reducing friction between the bit and the formation, which further enhances drilling efficiency [2, 3].

Drilling fluids also help to suspend cuttings and debris generated by the drilling process, preventing them from settling in the wellbore and interfering with drilling operations. The suspended cuttings are then transported to the surface by the drilling fluid, where they are separated from the fluid and disposed of. Furthermore, drilling fluids are used to control pressure in the wellbore, which is essential in preventing blowouts and other dangerous situations. The weight of the drilling fluid is carefully controlled to maintain a balance of pressure between the wellbore and the formation being drilled, preventing the formation fluids from entering the wellbore [4].

In conclusion, drilling fluids play a critical role in ensuring the safety and efficiency of drilling operations. They provide stability to the wellbore, cool and lubricate the drill bit, suspend cuttings and debris, and control pressure in the wellbore. The selection of the appropriate drilling fluid composition for a specific drilling operation is crucial, as it ensures that the drilling process is carried out safely and efficiently.

1.2 The need for innovation in drilling fluid design

The oil and gas industry is constantly evolving, and with it, the technology used in drilling operations. In recent years, there has been a growing need for innovation in drilling fluid design. As drilling operations become more complex, the demand for drilling fluids that can meet the specific requirements of each operation has increased. The following are some of the main reasons why innovation in drilling fluid design is crucial [5]:

i. Increasing complexity of drilling operations

As drilling operations become more complex, the requirements for drilling fluids have become more stringent. Today's drilling operations involve drilling in deeper waters, at higher pressures and temperatures, and through more challenging formations. This has led to a need for drilling fluids that can perform under extreme conditions, such as high-temperature and high-pressure environments, and can provide the necessary stability and lubrication required for efficient drilling.

ii. Environmental regulations

Environmental regulations have become increasingly strict in recent years, leading to a need for drilling fluids that are more environmentally friendly. Traditional drilling fluids contain toxic chemicals that can harm the environment and wildlife if they are released into the surrounding ecosystem. Therefore, there is a need for drilling fluids that are less toxic and biodegradable, which can reduce the impact of drilling operations on the environment.

iii. Cost-effective solutions

Drilling fluids can account for a significant portion of the total cost of drilling operations. As such, there is a need for cost-effective solutions that can meet the specific requirements of each drilling operation. Innovations in drilling fluid design can lead to the development of more efficient and cost-effective drilling fluids, which can help reduce the overall cost of drilling operations.

iv. Increasing demand for energy

The world's demand for energy is increasing, and the oil and gas industry is under pressure to meet this demand. This has led to an increase in drilling operations, and as such, a need for drilling fluids that can help increase drilling efficiency. Innovations in drilling fluid design can help develop fluids that can improve drilling efficiency, reduce the time required for drilling operations, and increase the overall output of oil and gas wells.

v. Safety and risk management

Drilling operations can be dangerous, and safety is always a top priority. Innovations in drilling fluid design can help develop fluids that are safer to handle, reducing the risk of accidents and injuries during drilling operations. Furthermore, drilling fluids that can provide better pressure control can reduce the risk of blowouts and other dangerous situations.

1.3 Insight into some tradition drilling fluid additives

Traditional drilling fluid additives have been used in the oil and gas industry for many years to optimize drilling operations. These additives are designed to improve the physical and chemical properties of the drilling fluid, which is crucial for efficient drilling and wellbore stability. Some of the commonly used traditional drilling fluid additives include [6]:

Bentonite clay: Bentonite is a natural clay mineral that is used as a viscosifier and filtration control agent in drilling fluids. It is effective in increasing the viscosity of the drilling fluid, which helps in maintaining the stability of the borehole. Bentonite clay also has good filtration control properties, which helps in preventing the loss of drilling fluid into the formation.

Barite: Barite is a mineral that is used as a weighting agent in drilling fluids. It is added to the drilling fluid to increase its density, which is necessary for controlling the formation pressure and preventing well blowouts. Barite also helps in maintaining the stability of the borehole by providing adequate hydrostatic pressure.

Caustic soda: Caustic soda (sodium hydroxide) is used as a pH control agent in drilling fluids. It helps in maintaining the alkalinity of the drilling fluid, which is important for preventing the corrosion of the drill string and other equipment. Caustic soda also helps in reducing the viscosity of the drilling fluid, which improves the efficiency of drilling.

Xanthan gum: Xanthan gum is a biopolymer that is used as a viscosifier in drilling fluids. It is effective in increasing the viscosity of the drilling fluid, which helps in maintaining the stability of the borehole. Xanthan gum is also shear-thinning, which means that its viscosity decreases under high shear conditions, such as when the drilling fluid is pumped through the drill bit.

Potassium chloride: Potassium chloride is used as a salt in drilling fluids. It is added to the drilling fluid to increase its density and provide sufficient hydrostatic pressure to control the formation pressure. Potassium chloride is also effective in stabilizing shale formations by inhibiting clay swelling.

These traditional drilling fluid additives have been used for many years and have proven to be effective in improving the properties of the drilling fluid. However, they also have some limitations and drawbacks. For example, some of these additives can be expensive and may have adverse environmental impacts. Additionally, they may not be effective in certain drilling conditions or formations. As a result, there is a growing interest in exploring alternative additives, such as deep eutectic solvents, that can provide comparable or superior performance while addressing some of these limitations.

2. Deep eutectic solvents

2.1 Definition of DES

Deep eutectic solvents (DESs) are a relatively new class of solvents that have been attracting increasing attention due to their unique properties and potential applications in various industries, including the oil and gas industry. DESs are typically formed by mixing two or more components, such as a hydrogen bond acceptor and a hydrogen bond donor, at a specific molar ratio, which results in a eutectic mixture that has a lower melting point than either of the individual components. The concept of eutectic mixtures has been known for over a century, but it was not until the early 21st century that the term 'deep eutectic solvents' was coined to describe these types of mixtures. The 'deep' part of the name refers to the fact that the melting point depression is significant and occurs at a temperature much lower than that of either of the individual components, which is not the case for traditional eutectic mixtures [7].

2.1.1 Eutectic mixtures

Eutectic mixtures are a type of mixture formed when two or more substances with different melting points are combined in certain proportions to form a new mixture that has a lower melting point than either of the individual components. The specific proportions at which the mixture will have the lowest melting point are called the eutectic point or eutectic composition. At this point, the mixture will solidify or melt at a single temperature instead of over a range of temperatures.

In the case of deep eutectic solvents (DESs), eutectic mixtures are formed between a hydrogen bond acceptor (HBA) and a hydrogen bond donor (HBD). These mixtures form a new solvent with unique properties such as low vapor pressure, low flammability, high thermal stability, and a wide range of tunable physical and chemical properties. The formation of a eutectic mixture is dependent on the molecular structure and properties of the HBA and HBD. The HBA typically has a higher melting point and stronger electron-withdrawing ability than the HBD. When combined in the correct proportions, the HBD can disrupt the lattice structure of the HBA, leading to a reduction in the melting point of the mixture. The formation of DES is a delicate balance between the strength of hydrogen bonding and the size and shape of the molecules involved. Too weak of a hydrogen bond or a mismatch in size and shape can result in a mixture that does not have the desired properties of a DES.

Overall, the formation of eutectic mixtures and subsequent development of DESs have opened up a new frontier in solvent design, with potential applications in a wide range of industries, including drilling fluid design.

2.2 Properties and characteristics of DES

Deep eutectic solvents (DESs) are a class of solvents that have been gaining increasing attention in various industries, including the oil and gas industry. Here are some of the key properties and characteristics of DES:

Low toxicity and biodegradability: DESs are often composed of natural compounds that are non-toxic and biodegradable, making them more environmentally friendly than traditional solvents.

Low vapor pressure: DES typically have low vapor pressures, which reduce the risk of volatile organic compounds (VOCs) being released into the atmosphere.

High thermal stability: DESs have high thermal stability, which allows them to be used in high-temperature applications without breaking down.

High solubility: DESs have high solubility for a wide range of organic and inorganic compounds, which makes them useful in various applications, including as drilling fluid additives.

Tunable properties: The properties of DES can be easily tuned by changing the composition of the solvent, allowing for customization to suit specific applications.

Non-flammability: DESs are typically non-flammable, which reduces the risk of fires and explosions in applications that involve high temperatures.

High viscosity: DESs often have higher viscosities than traditional solvents, which makes them useful as thickening agents or rheology modifiers in drilling fluids.

Good lubrication properties: DESs have good lubrication properties, which makes them useful in applications that involve friction or wear, such as drilling operations are free to decide how the main body will be structured. However, you are required to have at least one heading. Please ensure that either British or American English is used consistently in your chapter.

2.3 Utilization of DES in various fields

Deep eutectic solvents (DESs) are a type of ionic liquid that exhibit unique properties and have found diverse applications in various fields. Here are some details on the uses of DES and their mechanisms in different areas:

Green chemistry: DES can serve as environmentally friendly solvents for various chemical reactions, such as organic synthesis and catalysis [8–10].

Mechanism: DES provides a suitable medium for chemical reactions by dissolving reactants, stabilizing intermediates, and enhancing reaction rates through increased molecular mobility.

Extraction and separation: DES can be used as alternative solvent for the extraction and separation of organic compounds, metals, and biomolecules [11–13].

Mechanism: DESs can form strong interactions with the target compounds, leading to their selective extraction or separation from a mixture.

Electrochemistry: DES can act as electrolytes or electrode modifiers in electrochemical systems, including batteries, supercapacitors, and sensors [14, 15].

Mechanism: DES can facilitate ion transport, enhance electrode stability, and modulate the electrochemical behavior of the system.

Biocatalysis: DES can serve as reaction media for biocatalytic processes, enabling enzymatic reactions in non-aqueous or harsh conditions [16, 17].

Mechanism: DES can maintain the stability and activity of enzymes, improve substrate solubility, and enhance biocatalytic performance.

Pharmaceutical applications: DES can be utilized in drug delivery systems, formulation development, and synthesis of active pharmaceutical ingredients [18, 19].

Mechanism: DES can solubilize poorly soluble drugs, stabilize formulations, and provide a controlled release of therapeutic agents.

Material science: DES can be employed in the synthesis, processing, and modification of materials, including polymers, nanoparticles, and composites [20, 21].

Mechanism: DES can act as reaction media, dispersants, or template agents, influencing the morphology, properties, and self-assembly of materials.

The mechanisms underlying the properties and behavior of DES can vary depending on the specific composition and components involved. Generally, DESs operate through a combination of hydrogen bonding, ion-dipole interactions, and coordination between the components, leading to unique solvent characteristics and enabling their diverse applications.

3. Introduction to deep eutectic solvents and their potential as drilling fluid additives

In their pioneering work, Han Jia et al. (2019) investigated the use of deep eutectic solvents (DESs) as shale inhibitors. They tested three different DES compositions: Propoanoic acid ChCl (1:1), 3-phenyl propanoic acid ChCl (1:2), and a mixture of 3-mercapto propanoic acid, Itaconic acid, and ChCl (1:1:2). Their study revealed that these DES formulations provided 68%, 58%, and 58% inhibition of bentonite swelling, respectively [22]. In a subsequent study by M.H. Rasool et al. (2021), a Glycerine:Potassium Carbonate DES (2:1) was employed in a free-style experiment using swelling shale samples. Remarkably, this DES demonstrated 87% inhibition of swelling [23]. Moreover, M.H. Rasool et al. (2022) investigated a double-action approach using a Potassium carbonate-based DES in combination with a Poly (2-ethyl-2-Oxazoline) hydroxyl-terminated polymer in drilling mud. This innovative combination resulted in a remarkable 76% inhibition of swelling [24].

Deep eutectic solvents (DESs) have shown potential as drilling fluid additives for various applications, including mud rheology modification, shale swelling inhibition, and hydrate inhibition. A detailed discussion on each aspect has been carried out below.

3.1 As a rheology modifier

Drilling mud rheology refers to the study of the flow behavior of drilling fluids. Drilling mud is a critical component of the drilling process, used to cool and lubricate the drill bit, carry drill cuttings to the surface, and maintain pressure in the wellbore. The flow properties of the drilling mud, including viscosity, yield point, gel strength, and fluid loss, are crucial to ensure efficient and safe drilling operations. Viscosity is the most fundamental property of drilling mud rheology and refers to the fluid's resistance to flow. The viscosity of the drilling mud must be carefully controlled to prevent excessive pressure buildup, which can cause wellbore instability, lost circulation, and other drilling problems. Yield point is another important rheological property, representing the minimum stress required to initiate fluid movement. Gel strength measures the shear strength of the drilling mud, and fluid loss measures the amount of fluid that is lost to the formation during drilling operations.

Deep eutectic solvents (DESs) have shown potential as drilling fluid additives for various applications, including mud rheology modification. One of the most common components of drilling fluids is sodium bentonite, which is a clay mineral that provides viscosity and filtration control properties to the drilling fluid [25]. The interaction between DES and sodium bentonite in drilling mud can improve the rheology of the fluid and provide other benefits. The interaction between DES and sodium bentonite is believed to involve the formation of hydrogen bonds between the HBA and HBD components of DES and the hydroxyl groups of sodium bentonite. The hydrogen bonding can lead to changes in the interlayer spacing of the sodium bentonite and improve the rheological properties of the drilling mud. DES can act as both a thickener and a thinner of drilling fluids depending on the HBA and HBD components used [26].

For example, DES based on choline chloride and urea can increase the viscosity of sodium bentonite-based drilling fluids. The interaction between DES and sodium bentonite can cause an increase in the interlayer spacing of sodium bentonite, leading to an increase in the viscosity of the drilling fluid. In addition, DES can reduce the gel strength of the drilling mud, which can improve the flow properties of the fluid and reduce the risk of differential sticking. On the other hand, DES based on choline chloride and ethylene glycol can decrease the viscosity of sodium bentonite-based drilling fluids. The interaction between DES and sodium bentonite can cause a decrease in the interlayer spacing of sodium bentonite, leading to a decrease in the viscosity of the drilling fluid. This can be useful in situations where a lower-viscosity drilling fluid is required. Furthermore, DES can also improve the thermal stability of sodium bentonite-based drilling fluids. The interaction between DES and sodium bentonite can help to prevent thermal degradation of the fluid and improve the thermal stability of the fluid. However, there are still challenges that need to be addressed when using DES as drilling fluid additives, including issues related to the compatibility of DES with other drilling fluid components and the need for further research to fully understand the behavior of DES under different drilling conditions.

In conclusion, the interaction between DES and sodium bentonite in drilling mud can improve the rheology of the fluid and provide other benefits. The mechanism of DES interaction with sodium bentonite is believed to involve the formation of hydrogen bonds between the HBA and HBD components of DES and the hydroxyl groups of sodium bentonite. DES can act as both a thickener and a thinner of drilling fluids depending on the HBA and HBD components used. Further research is needed to optimize the use of DES in drilling fluids and address compatibility issues. DES can also be used as shale swelling inhibitors in drilling fluids. Shale swelling is a common issue in drilling operations, which can lead to significant challenges, including reduced drilling efficiency and wellbore instability. DESs can inhibit shale swelling by penetrating the shale formation and interacting with the clay minerals, preventing them from absorbing water and swelling. The mechanism of DES as a shale swelling inhibitor is believed to involve the disruption of the hydrogen bonding between the clay minerals and water molecules. This can prevent the swelling of the shale formation and improve the stability of the wellbore.

3.2 As a hydrate inhibitor

During deepwater drilling operations, the high-pressure and low-temperature conditions can cause the formation of gas hydrates, which are crystalline structures composed of water and gas molecules. These hydrates can cause significant problems in drilling operations, such as plugging of equipment and pipes, reducing the efficiency of the drilling process, and even leading to safety hazards. Gas hydrates form when gas molecules (typically methane, ethane, propane, and butane) are trapped within the lattice structure of water molecules, under conditions of high pressure and low temperature. In deepwater drilling, the temperature can drop below the hydrate formation temperature of the gas, and the pressure can be high enough to stabilize the hydrate structure. Once formed, these hydrates can accumulate in pipes, valves, and other equipment, reducing the flow of drilling fluid and increasing the risk of blockages.

Deep eutectic solvents (DESs) have shown potential as drilling fluid additives for various applications, including the inhibition of hydrate formation during deepwater drilling operations. DES can act as hydrate inhibitors by forming a hydrate-inhibiting film on the surface of gas hydrate crystals, which prevents further growth and agglomeration of the crystals as shown in **Figure 1**. The mechanism of DES inhibition of hydrate formation involves a combination of thermodynamic and kinetic effects.

The thermodynamic effect is due to the ability of DES to reduce the driving force for hydrate formation by lowering the equilibrium temperature and pressure for hydrate formation. This is because DES can dissolve in water and reduce the concentration of free water molecules, which are required for hydrate formation. As a result,

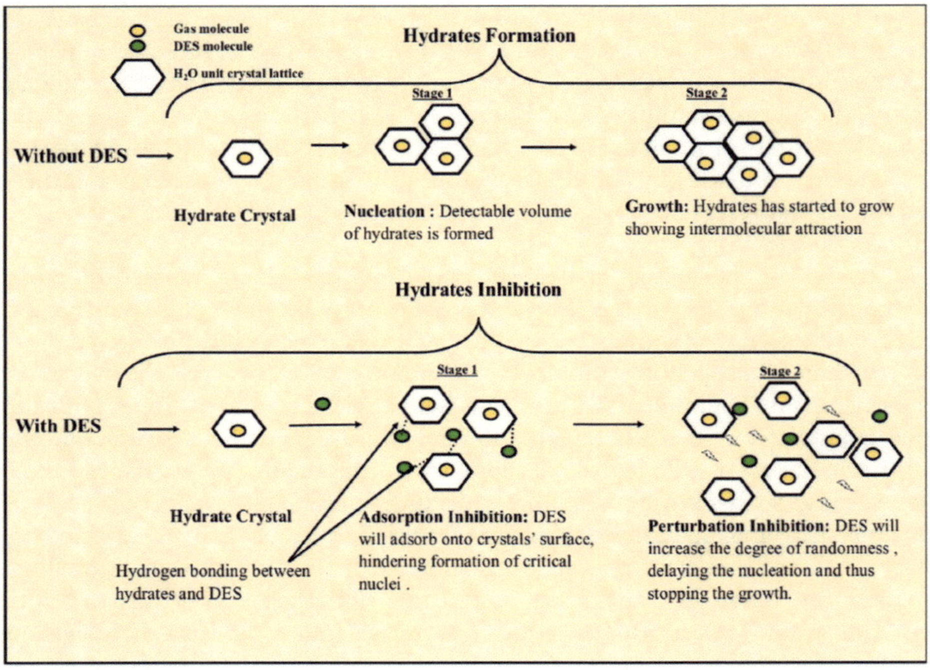

Figure 1.
DES as a hydrate inhibitor [27].

the formation of hydrates is suppressed, and the stability of existing hydrate crystals is reduced. The kinetic effect is due to the ability of DES to adsorb onto the surface of hydrate crystals and form a film that prevents further growth and agglomeration of the crystals. The adsorption of DES onto the surface of hydrate crystals is believed to be facilitated by the presence of hydrogen bonding sites on the surface of the crystals. The hydrogen bonding between the HBA and HBD components of DES and the hydrogen bonding sites on the surface of hydrate crystals can lead to the formation of a stable film that inhibits further growth of the crystals.

One of the most promising DES for hydrate inhibition is based on choline chloride and urea. This DES has been shown to be effective in inhibiting hydrate formation in laboratory experiments and in field trials. In addition, this DES has the advantage of being environmentally friendly and biodegradable, which is important for reducing the environmental impact of drilling operations. However, there are still challenges that need to be addressed when using DES as hydrate inhibitors in drilling operations, including issues related to the compatibility of DES with other drilling fluid components and the need for further research to fully understand the behavior of DES under different drilling conditions.

In conclusion, DES has the potential to be effective hydrate inhibitor in deep-water drilling operations. The mechanism of DES inhibition of hydrate formation involves a combination of thermodynamic and kinetic effects, which reduce the driving force for hydrate formation and prevent further growth and agglomeration of hydrate crystals. Further research is needed to optimize the use of DES as hydrate inhibitors in drilling fluids and address compatibility issues.

3.3 As a shale inhibitor

Shale is a fine-grained sedimentary rock composed of clay minerals such as illite, kaolinite, and smectite. Shale is a common rock encountered in oil and gas drilling, and it poses several challenges to drilling operations due to its unique properties. One of the major issues with shale is its tendency to swell when exposed to water-based drilling fluids. This phenomenon is known as shale swelling, and it can result in significant problems such as stuck pipe, lost circulation, and formation damage. Shale swelling is primarily caused by the interaction between the clay minerals in shale and water-based drilling fluids. When shale is exposed to water-based drilling fluids, the clay minerals in the shale absorb water and begin to swell. This swelling can cause the shale to expand and reduce the permeability of the formation, making it difficult for the drilling fluid to circulate through the wellbore [25, 28].

Deep eutectic solvents (DESs) have shown potential as drilling fluid additives for inhibiting shale swelling during drilling operations. The mechanism of DES inhibition of shale swelling involves a combination of physical and chemical effects. The physical effect is due to the ability of DES to reduce the hydration of clay minerals in shale formations. Clay minerals have a layered structure that consists of negatively charged sheets of silica and alumina, with positively charged ions such as sodium and potassium located between the layers. When water is present, the positively charged ions interact with the negatively charged sheets, causing the layers to swell and expand. DES can interact with these positively charged ions, reducing the amount of water that can access the clay mineral layers and reducing the amount of swelling as shown in **Figure 2** [7, 29]. The chemical effect is due to the ability of DES to interact with the clay minerals at a molecular level. DES can penetrate into the interlayer space

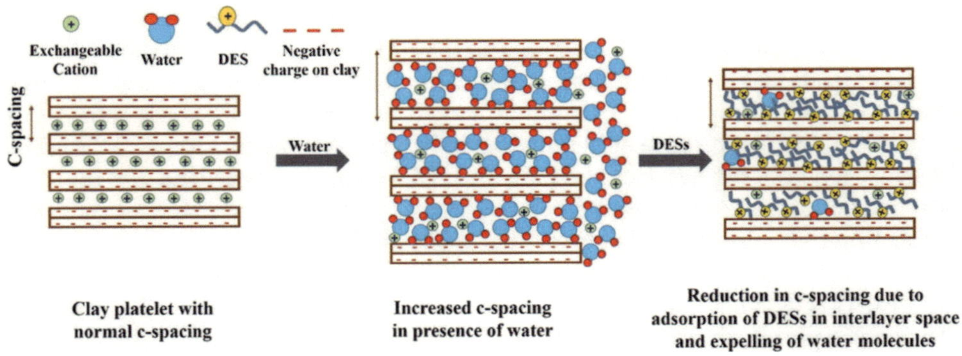

Figure 2.
DES as a shale inhibitor [7].

of clay minerals and disrupt the interactions between the layers. This can lead to the delamination of the clay minerals, reducing their ability to swell and expand.

One of the most promising DES for shale swelling inhibition is based on choline chloride and urea. This DES has been shown to be effective in reducing shale swelling in laboratory experiments and in field trials. In addition, this DES has the advantage of being environmentally friendly and biodegradable, which is important for reducing the environmental impact of drilling operations. However, there are still challenges that need to be addressed when using DES as shale swelling inhibitors in drilling operations, including issues related to the compatibility of DES with other drilling fluid components and the need for further research to fully understand the behavior of DES under different drilling conditions [30, 31].

In conclusion, DESs have the potential to be used as drilling fluid additive for various applications, including mud rheology modification, shale swelling inhibition, and hydrate inhibition. The mechanism of DES in each application is believed to involve specific interactions with the components of the drilling fluid and the geological formations. Further research is needed to fully understand the behavior of DES under different drilling conditions and to optimize their use in drilling fluids.

3.4 Advantages and disadvantages of using DES as drilling fluid additives

Deep eutectic solvents (DESs) have gained attention as potential additives for drilling fluids due to their unique properties. However, as with any new technology, there are advantages and disadvantages to their use in drilling operations [23, 32, 33].

3.4.1 Advantages of using DES as drilling fluid additives

Environmentally friendly: DESs are typically composed of natural compounds that are non-toxic and biodegradable, making them a more environmentally friendly alternative to traditional solvents.

Low toxicity: DESs have been shown to have low toxicity, which can reduce the risk of negative health effects for workers who come into contact with the drilling fluid.

High solubility: DESs have high solubility for a wide range of organic and inorganic compounds, making them useful as a means of controlling drilling fluid properties.

Customizable: The properties of DES can be easily tuned by changing the composition of the solvent, allowing for customization to suit specific drilling applications.

High thermal stability: DESs have high thermal stability, which allows them to be used in high-temperature applications without breaking down.

3.4.2 Disadvantages of using DES as drilling fluid additives

Limited research: DESs are a relatively new technology and there is limited research on their long-term effectiveness and environmental impact.

Limited compatibility: DESs may not be compatible with all types of drilling fluids, which could limit their use in certain applications.

Limited availability: DESs may not be readily available in all regions, which could limit their use in certain applications.

Potential for increased viscosity: DES can have high viscosity, which could lead to increased pressure drop and potential problems with equipment [24].

In conclusion, DESs have unique advantages as drilling fluid additives due to their environmental friendliness, high solubility, and tunable properties. However, their limited research and higher cost, as well as potential compatibility issues and increased viscosity, may limit their use in certain drilling applications. Further research is needed to fully understand the benefits and limitations of DES as drilling fluid additives.

3.5 Potential combination of DES as drilling fluid additives

There are numerous combinations of deep eutectic solvents (DESs) that have been successfully used as drilling fluid additives. Some of these include [22, 25, 34, 35]:

Choline chloride and urea – This combination has been shown to effectively inhibit shale swelling and improve the thermal stability of the drilling fluid.

Choline chloride and glycerol – This combination has been used as a lubricant in drilling fluids and has shown to be effective in reducing the coefficient of friction.

Choline chloride and ethylene glycol – This combination has been shown to have excellent hydrate inhibition properties and can be used in deep water drilling applications.

Choline chloride and lactic acid – This combination has been used as a pH buffer and has been shown to improve the stability of the drilling fluid.

Betaine and ethylene glycol – This combination has been shown to have good hydrate inhibition properties and can be used in deep water drilling applications.

Betaine and glycerol – This combination has been used as a lubricant in drilling fluids and has shown to be effective in reducing the coefficient of friction.

Betaine and urea – This combination has been shown to effectively inhibit shale swelling and improve the thermal stability of the drilling fluid.

Betaine and lactic acid – This combination has been used as a pH buffer and has been shown to improve the stability of the drilling fluid.

Levulinic acid and dimethyl sulfoxide – (DMSO) as a deep eutectic solvent for improving the thermal stability of water-based drilling fluids at high temperatures.

Eucalyptol and 1,3-propanediol as a deep eutectic solvent for improving the lubricity of oil-based drilling fluids.

N-methyl-2-pyrrolidone (NMP) and choline chloride as a deep eutectic solvent for improving the wettability of shale surfaces and preventing shale hydration.

Glycerol and malonic acid as a deep eutectic solvent for inhibiting the formation of gas hydrates in deepwater drilling operations.

Imidazole and choline chloride as a deep eutectic solvent for reducing the fluid loss of oil-based drilling fluids and preventing differential sticking.

Glucose and levulinic acid as a deep eutectic solvent for improving the fluid loss control and filtration properties of water-based drilling fluids.

N-methylacetamide and choline chloride as a deep eutectic solvent for enhancing the thermal stability and rheological properties of water-based drilling fluids at high temperatures and pressures.

Overall, the use of DES as drilling fluid additive has shown great promise in improving the performance and efficiency of drilling operations. As research continues, it is likely that new combinations of DESs will be developed and tested for their potential as drilling fluid additives.

3.6 Ionic liquids vs. DES as drilling fluid additives

Ionic liquids and deep eutectic solvents (DESs) are both types of solvents that have gained attention for their potential applications in various fields, including drilling fluid additives. While both are considered "green solvents" due to their low toxicity and biodegradability, there are key differences between them. Ionic liquids (ILs) are salts that are liquid at room temperature, typically composed of large, asymmetric cations and small, inorganic anions. The ionic nature of these compounds means that they have a high degree of polarity and can dissolve a wide range of materials, including inorganic salts, polymers, and gases. However, their high cost and potential toxicity have limited their widespread use [36].

In contrast, DESs are formed through the interaction of two or more solid components, typically a hydrogen bond acceptor (HBA) and a hydrogen bond donor (HBD), that undergo a eutectic reaction to form a liquid. This results in a solvent that has a lower melting point than either of its individual components and exhibits unique properties such as high polarity, low viscosity, and tunable solubility. DES can be composed of a variety of HBAs and HBDs, allowing for a wide range of possible solvents with different properties. One of the main advantages of DESs over ILs is their lower cost, as the solid components used to form DESs are typically more readily available and less expensive than the chemicals used to produce ILs. Additionally, DESs are generally less toxic and have a lower environmental impact than ILs.

In terms of their applications as drilling fluid additives, both ILs and DESs have shown promise in improving drilling fluid properties such as viscosity, lubricity, and stability. However, DESs are considered to be more suitable for use in this field due to their lower cost, lower toxicity, and compatibility with other drilling fluid additives. In summary, while both Ils and DES are considered "green solvents" with potential applications in a variety of fields, including drilling fluid design, DESs have certain advantages over ILs in terms of cost, toxicity, and compatibility with other additives.

4. Environmental benefits of DES-based drilling fluids

Deep eutectic solvents (DESs) are an emerging class of solvents that have gained attention as potential drilling fluid additives due to their unique properties.

In addition to their tunable properties and high solubility, DES-based drilling fluids also offer several environmental benefits that make them an attractive alternative to traditional drilling fluids [37].

Biodegradability: DESs are typically composed of natural compounds that are biodegradable, meaning that they break down into harmless substances in the environment. This is in contrast to traditional drilling fluids, which can contain toxic chemicals that can persist in the environment for long periods of time.

Reduced toxicity: DESs have been shown to have low toxicity, which can reduce the risk of negative health effects for workers who come into contact with the drilling fluid. This can also reduce the risk of harm to wildlife and ecosystems in the event of spills or leaks.

Reduced environmental impact: The use of DES-based drilling fluids can help to reduce the environmental impact of drilling operations by reducing the amount of toxic chemicals that are released into the environment. This can help to protect water sources, wildlife, and ecosystems in the surrounding areas.

Energy efficiency: DESs have a lower vapor pressure than traditional solvents, which means that they require less energy to maintain their liquid state. This can help to reduce the energy consumption and associated greenhouse gas emissions of drilling operations.

Renewable resources: Many DESs are derived from renewable resources such as plants, which means that they can be produced in a sustainable and environmentally friendly manner.

Overall, DES-based drilling fluids offer several environmental benefits over traditional drilling fluids. By reducing the environmental impact of drilling operations, these fluids can help to protect the environment and promote sustainable drilling practices.

5. Compatibility with other drilling fluid additives

The compatibility of deep eutectic solvents (DESs) with other drilling fluid additives is an important factor in determining their overall effectiveness in drilling operations. Here we will discuss the compatibility of DES with various additives commonly used in drilling fluids, including soda ash, caustic soda, xanthan gum, barite, KCl, and cellulose starch.

Soda ash: Soda ash can be used to remove hardness in water-based drilling fluids. Hardness in water is caused by the presence of dissolved calcium and magnesium ions, which can negatively affect the performance of drilling fluids by causing the precipitation of certain additives and reducing their effectiveness. Soda ash reacts with the calcium and magnesium ions in the water to form insoluble carbonates, which can be easily removed through filtration or settling. This process is known as water softening and can improve the performance of drilling fluids by reducing the amount of scale buildup and improving the efficiency of certain additives.

Caustic soda: Caustic soda, or sodium hydroxide, is used in drilling fluids to adjust pH levels, as well as to inhibit the swelling of clays. DES can be compatible with caustic soda, but again, it is important to ensure that the pH is within the desired range for both additives. In addition, caustic soda can increase the salt content of the drilling fluid, which can impact the solubility of the DES.

Xanthan gum: Xanthan gum is a commonly used viscosity modifier in drilling fluids. DES can be compatible with xanthan gum, but it is important to ensure that the concentrations of both additives are within acceptable ranges. High concentrations of xanthan gum can lead to decreased effectiveness of the DES, as well as potential precipitation or separation of the DES from the fluid.

Barite: Barite is a weighting agent used in drilling fluids to increase density. DES can be compatible with barite, but it is important to ensure that the concentrations of both additives are within acceptable ranges. High concentrations of barite can lead to decreased effectiveness of the DES, as well as potential precipitation or separation of the DES from the fluid.

KCl: KCl, or potassium chloride, is used in drilling fluids as a shale inhibitor and to increase fluid density. DES can be compatible with KCl, but again, it is important to ensure that the concentrations of both additives are within acceptable ranges. High concentrations of KCl can lead to decreased effectiveness of the DES, as well as potential precipitation or separation of the DES from the fluid.

Cellulose starch: Cellulose starch is a commonly used viscosifier in drilling fluids. DES can be compatible with cellulose starch, but it is important to ensure that the concentrations of both additives are within acceptable ranges. High concentrations of cellulose starch can lead to decreased effectiveness of the DES, as well as potential precipitation or separation of the DES from the fluid.

In general, the compatibility of DES with other drilling fluid additives is dependent on the specific composition of the DES and the other additives used. It is important to carefully consider the concentrations and pH levels of all additives in the drilling fluid, and to test the compatibility of the DES with other additives before using it in drilling operations.

6. Challenges and future outlook

6.1 Use of DES in high-temperature and high-pressure drilling environments

Deep eutectic solvents (DESs) have shown great potential as drilling fluid additives in high-temperature and high-pressure drilling environments. These conditions are often encountered in deepwater and ultra-deepwater drilling, where temperatures and pressures can exceed 150°C and 20,000 psi, respectively. Traditional drilling fluids may not be able to withstand these extreme conditions and can break down, resulting in decreased drilling efficiency and safety hazards. One of the key advantages of using DESs in high-temperature and high-pressure drilling environments is their excellent thermal stability. DESs have been shown to remain stable at temperatures up to 200°C and pressures up to 1000 bar. This makes them ideal for use in deepwater and ultra-deepwater drilling applications, where temperatures and pressures are at their highest. DESs also have good lubricity and can reduce the coefficient of friction between the drill string and the borehole wall. This can help to reduce wear and tear on the drill string and improve drilling efficiency [38].

6.2 Cost-effectiveness of DES-based drilling fluids

Deep eutectic solvents (DESs) have gained attention as a potential alternative to conventional drilling fluid additives due to their unique properties and potential advantages. One advantage that has been explored is the cost-effectiveness of using DES-based drilling fluids. The cost-effectiveness of drilling fluids is an important factor to consider in the oil and gas industry, as the cost of drilling can be a significant expense. DES-based drilling fluids have the potential to be more cost-effective than traditional drilling fluids for several reasons.

First, DES can be synthesized from relatively inexpensive starting materials, such as choline chloride and urea. This means that the cost of producing DES can be lower than that of traditional drilling fluid additives. Additionally, DES can be used at lower concentrations than traditional additives, further reducing the cost of the fluid. DES can also potentially reduce the overall cost of drilling by improving drilling efficiency. The unique properties of DES can improve drilling performance by reducing friction between the drill bit and the formation being drilled. This can lead to faster drilling times and less wear on the drilling equipment, reducing maintenance costs.

Another potential cost-saving benefit of DES-based drilling fluids is their ability to be reused. Traditional drilling fluids can become contaminated with rock cuttings and other debris, which can render them unusable and require disposal. DES-based fluids, however, have been shown to be more stable and less prone to contamination, making them potentially reusable and reducing the need for frequent fluid changes. Overall, the cost-effectiveness of DES-based drilling fluids is a promising advantage that warrants further exploration and development. As research into the use of DES in drilling fluids continues, it is likely that more cost-saving benefits will be identified [39].

6.3 Potential future applications of DES in drilling fluid design

Deep eutectic solvents (DESs) have shown promising results as drilling fluid additives due to their ability to modify the rheology of the drilling fluid, inhibit shale swelling, and prevent hydrate formation. However, the potential applications of DES in drilling fluid design go beyond these uses.

One potential application is in enhancing the lubricity of the drilling fluid. Lubricity is a critical property for reducing friction and wears on the drill bit and the drilling equipment. DESs have been found to exhibit good lubricating properties, and their use in drilling fluid design could lead to improved drilling efficiency and reduced wear on equipment. Another potential application of DES is in enhancing the thermal stability of the drilling fluid. In high-temperature and high-pressure drilling environments, the drilling fluid is subjected to extreme conditions that can cause thermal degradation and loss of viscosity. DESs have been shown to have good thermal stability, and their incorporation into the drilling fluid could improve the fluid's performance in such environments [39].

DESs also have the potential to act as surfactants in drilling fluids, which could enhance their ability to wet and disperse solids and improve their emulsifying properties. This could lead to improved suspension of drill cuttings and better removal of solids from the wellbore. In addition, DESs have been investigated for their potential as environmentally friendly drilling fluid additives. They are biodegradable and have low toxicity, which makes them an attractive alternative to conventional drilling fluid additives that can have harmful environmental impacts.

Overall, DESs have shown great potential as versatile and innovative drilling fluid additives with a wide range of potential applications beyond their current uses. Further research and development are needed to fully explore their capabilities and optimize their performance in drilling operations [40].

6.4 Challenges and limitations

Despite their potential benefits, there are still several challenges and limitations associated with the use of deep eutectic solvents (DESs) as drilling fluid additives. Some of the major challenges are discussed below:

Limited availability: The production of DESs on a commercial scale is still limited, which makes them relatively unsuitable compared to other drilling fluid additives.

Limited understanding of DES behavior: There is still limited understanding of the behavior of DESs under various drilling conditions, which can make it difficult to optimize their use as drilling fluid additives.

Compatibility issues: Although DESs have shown compatibility with many drilling fluid additives, there are still some additives with which DES may not be compatible, which can limit their use in certain applications.

Stability at very high temperatures: Some DES may not be stable at very high temperatures, which can limit their use in high-temperature drilling environments.

Environmental concerns: While DESs have been touted as more environmentally friendly alternatives to traditional drilling fluid additives, there are still concerns about their long-term impact on the environment.

Health and safety concerns: The use of DES as drilling fluid additives may pose certain health and safety risks to workers in the drilling industry. It is important to properly handle and dispose of DES to minimize these risks though risks associated will be less severe than those posed by ionic liquids and oil-based mud.

Regulatory approval: The use of DES as drilling fluid additives may require regulatory approval in some jurisdictions, which can limit their use in certain applications.

In conclusion, while DESs offer many potential benefits as drilling fluid additives, there are still several challenges and limitations that need to be addressed before they can be widely adopted in the drilling industry. Further research is needed to better understand their behavior and optimize their use under various drilling conditions.

7. Conclusion

In conclusion, deep eutectic solvents (DESs) are a promising alternative to traditional drilling fluid additives due to their unique properties and potential benefits. They can serve as mud rheology modifiers, shale swelling inhibitors, and hydrate inhibitors during drilling operations. DES can improve the rheological properties of drilling mud by interacting with sodium bentonite, which increases the stability of the fluid and prevents solids from settling. DES can also inhibit hydrate formation during deep-water drilling by disrupting the formation of hydrate crystals. Furthermore, they can act as shale swelling inhibitors by coating the shale surface and preventing the absorption of water.

However, there are some limitations and challenges associated with the use of DES as drilling fluid additives, including their high cost and potential environmental impact. Compatibility issues with certain drilling formations and other additives must also be considered. Health and safety concerns should also be taken into account when handling DES. Despite these challenges, DES has shown potential in improving the efficiency and effectiveness of drilling operations. Further research is needed to explore the potential of DES-based drilling fluids in high-temperature and high-pressure drilling environments, as well as their compatibility with a wider range of drilling formations and additives. As the oil and gas industry continues to evolve and adapt to new environmental and regulatory requirements, DES may prove to be a valuable tool for sustainable and effective drilling fluid design.

When you are citing sources, the citations should be set in numbered format. All the references given in the list of references should be cited in the body of the text. Please set citations in square brackets keeping the below points in mind.

Acknowledgements

The authors would like to thank YUTP grant 015LC0-326 for providing the support.

Conflict of interest

The authors do not have any conflict of interest to disclose.

Author details

Muhammad Hammad Rasool* and Maqsood Ahmad
Petrloleum Geosciences Department, Universiti Teknologi Petronas, Malaysia

*Address all correspondence to: muhammad_19000949@utp.edu.my

References

[1] Hao Y. Research and application of anti-leakage drilling fluid. Bulgarian Chemical Communications. 2016;**48**:215-221

[2] Kårstad E, Aadnøy BS. Density behavior of drilling fluids during high pressure high temperature drilling operations. In: IADC/SPE Asia Pacific Drilling Technology. Kualalumpur, Malaysia. OnePetro; 1998

[3] Ravi A. Experimental Assessment of Water Based Drilling Fluids in High Pressure and High Temperature Conditions. Doha, Qatar: Texas A & M University; 2012

[4] Wójcik M, Kostowski W. Environmental risk assessment for exploration and extraction processes of unconventional hydrocarbon deposits of shale gas and tight gas: Pomeranian and Carpathian region case study as largest onshore oilfields. Journal of Earth Science. 2020;**31**(1):215-222

[5] Wang C-Q et al. Human carcinogenic risk analysis and utilization of shale gas water-based drilling cuttings in road materials. Environmental Science and Pollution Research. 2023;**30**(5):12741-12768

[6] Ali I, Ahmad M, Ganat T. Biopolymeric formulations for filtrate control applications in water-based drilling muds: A review. Journal of Petroleum Science and Engineering. 2022;**210**:110021

[7] Smith EL, Abbott AP, Ryder KS. Deep eutectic solvents (DESs) and their applications. Chemical Reviews. 2014;**114**(21):11060-11082

[8] Chang XX et al. A review on the properties and applications of chitosan, cellulose and deep eutectic solvent in green chemistry. Journal of Industrial and Engineering Chemistry. 2021;**104**:362-380

[9] Durand E, Lecomte J, Villeneuve P. From green chemistry to nature: The versatile role of low transition temperature mixtures. Biochimie. 2016;**120**:119-123

[10] Calvo-Flores FG, Mingorance-Sánchez C. Deep eutectic solvents and multicomponent reactions: Two convergent items to green chemistry strategies. ChemistryOpen. 2021;**10**(8):815-829

[11] Li X, Row KH. Development of deep eutectic solvents applied in extraction and separation. Journal of Separation Science. 2016;**39**(18):3505-3520

[12] Huang J, Guo X, Xu T, Fan L, Zhou X, Wu S. Ionic deep eutectic solvents for the extraction and separation of natural products. Journal of Chromatography A. 2019;**1598**:1-19

[13] Yucui H, Congfei Y, Weize W. Deep eutectic solvents: Green solvents for separation applications. Acta Physico-Chimica Sinica. 2018;**34**(8):873-885

[14] Abbott AP. Deep eutectic solvents and their application in electrochemistry. Current Opinion in Green and Sustainable Chemistry. 2022;**36**:100649

[15] Nkuku CA, LeSuer RJ. Electrochemistry in deep eutectic solvents. The Journal of Physical Chemistry B. 2007;**111**(46):13271-13277

[16] Xu P, Zheng G-W, Zong M-H, Li N, Lou W-Y. Recent progress on

deep eutectic solvents in biocatalysis. Bioresources and Bioprocessing. 2017;**4**:1-18

[17] Pätzold M, Siebenhaller S, Kara S, Liese A, Syldatk C, Holtmann D. Deep eutectic solvents as efficient solvents in biocatalysis. Trends in Biotechnology. 2019;**37**(9):943-959

[18] Emami S, Shayanfar A. Deep eutectic solvents for pharmaceutical formulation and drug delivery applications. Pharmaceutical Development and Technology. 2020;**25**(7):779-796

[19] Pedro SN, Freire CS, Silvestre AJ, Freire MG. Deep eutectic solvents and pharmaceuticals. Encyclopedia. 2021;**1**(3):942-963

[20] Nahar Y, Thickett SC. Greener, faster, stronger: The benefits of deep eutectic solvents in polymer and materials science. Polymers. 2021;**13**(3):447

[21] Ge X, Gu C, Wang X, Tu J. Deep eutectic solvents (DESs)-derived advanced functional materials for energy and environmental applications: Challenges, opportunities, and future vision. Journal of Materials Chemistry A. 2017;**5**(18):8209-8229

[22] Jia H et al. Investigation of inhibition mechanism of three deep eutectic solvents as potential shale inhibitors in water-based drilling fluids. Fuel. 2019;**244**:403-411

[23] Rasool MH, Zamir A, Elraies KA, Ahmad M, Ayoub M, Abbas MA. Potassium carbonate based deep eutectic solvent (DES) as a potential drilling fluid additive in deep water drilling applications. Petroleum Science and Technology. 2021;**39**(15-16):612-631

[24] Rasool MH, Ahmad M, Abbas MA. A double action PD (polymer-deep eutectic solvent) based shale inhibitor in drilling mud. Journal of Advanced Research in Fluid Mechanics and Thermal Sciences. 2022;**99**(1):149-157

[25] Ma J, Pang S, Zhou W, Xia B, An Y. Novel deep eutectic solvents for stabilizing clay and inhibiting shale hydration. Energy & Fuels. 2021;**35**(9):7833-7843

[26] Ma J, Pang S, An Y. Deep eutectic solvents for enhancing the rheological behavior of polymers and clays in polymeric water-based drilling fluids. Energy & Fuels. 2023;**37**(6):4391-4400

[27] Rasool MH, Zamir A, Elraies KA, Ahmad M, Ayoub M, Abbas MA. A deep eutectic solvent based novel drilling mud with modified rheology for hydrates inhibition in deep water drilling. Journal of Petroleum Science and Engineering. 2022;**211**:110151

[28] Jia H et al. Study of a gemini surface active ionic liquid 1, 2-bis (3-hexylimidazolium-1-yl) ethane bromide as a high performance shale inhibitor and inhibition mechanism. Journal of Molecular Liquids. 2020;**301**:112401

[29] Beg M, Haider MB, Thakur NK, Husein M, Sharma S, Kumar R. Clay-water interaction inhibition using amine and glycol-based deep eutectic solvents for efficient drilling of shale formations. Journal of Molecular Liquids. 2021;**340**:117134. DOI: 10.1016/j.molliq.2021.117134

[30] Jia H et al. Investigation of anionic group effects on the shale inhibition performance of fatty acid-based ionic liquids and their inhibition mechanism. Colloids and Surfaces A: Physicochemical and Engineering Aspects. 2022;**636**:128135

[31] Sivabalan V, Sahith JK, Lal B. Deep eutectic solvents as the new norm for oil and gas industry: A mini review. In: Third International Conference on Separation Technology 2020 (ICoST 2020). Amsterdam, Netherlands: Atlantis Press; 2020. pp. 119-124

[32] Rasool MH, Ahmad M, Ayoub M, Zamir A, Abbas MA. A review of the usage of deep eutectic solvents as shale inhibitors in drilling mud. Journal of Molecular Liquids. 2022;**361**:119673

[33] Rasool MH, Ahmad M, Ayoub M, Abbas MA. A novel ascorbic acid based natural deep eutectic solvent as a drilling mud additive for shale stabilization. PRO. 2023;**11**(4):1135

[34] Sultana K, Rahman MT, Habib K, Das L. Recent advances in deep eutectic solvents as shale swelling inhibitors: A comprehensive review. ACS Omega. 2022;**7**(33):28723-28755

[35] Muhammad MHR. Deep Eutectic Solvent Based Novel Drilling Mud with Improved Rheological and Filtration Behaviour. Bandar Seri Iskander, Malaysia: Universiti Teknologi PETRONAS; 2021

[36] Sanati A, Malayeri MR, Busse O, Weigand JJ. Utilization of ionic liquids and deep eutectic solvents in oil operations: Progress and challenges. Journal of Molecular Liquids. 2022;**361**:119641

[37] Bai J et al. Investigation of the mechanism and effect of citric acid-based deep eutectic solvents inhibiting hydration and expansion of gas shale clay minerals. Energy & Fuels. 2023;**37**(4):2750-2758

[38] Padinhattath SP, Chenthamara B, Gardas RL. Ionic liquids as alternative solvents for energy conservation and environmental engineering. Acta Innovations. 2021;**38**:62-79

[39] Ma J, Xu J, Pang S, Zhou W, Xia B, An Y. Novel environmentally friendly lubricants for drilling fluids applied in shale formation. Energy & Fuels. 2021;**35**(9):8153-8162

[40] Beg M, Kesarwani H, Sharma S, Saxena A. Impact of low-molecular-weight poly (4-styrenesulfonic acid-co-maleic acid) sodium salt on filtration and rheological parameters of nanoparticles-enhanced drilling fluid. Journal of Vinyl and Additive Technology. 2022;**28**(1):125-139

Chapter 4

1D Modeling of CO_2 Absorption Using K_2CO_3 in a Hollow Fibre Membrane Contactor

Mohamed Nadir Khelifi, Ouacil Saouli, Anis Bouzeraib and Khaled Basta

Abstract

Carbon dioxide (CO_2) is widely recognized as the main cause of climate change affecting our planet. In recent years, the concentration of CO_2 in the atmosphere has increased significantly due to the intensive combustion of fossil fuels. This study is devoted to the simulation of the 1D case of the CO_2 capture process by aqueous Potassium carbonate K_2CO_3 as a chemical solvent, using a membrane contactor in the counter current case. The system of partial differential equations resulting from the modeling was solved using MATLAB's PDEPE function. We also carried out a parametric analysis to see the impact of various parameters on the CO_2 capture process. Among these parameters, we studied the influence of solvent concentration, gas velocity and liquid velocity. The results show that 13% increase (74–87%) in CO_2 capture while the increase of solvent concentration from 20 to 50 (mol/m^3), also for gas and liquid velocity from 0.001 to 0.05 and 0.006 to 0.05 (m/s) we have 52% and 3% increase of CO_2 capture respectively.

Keywords: absorption, CO_2, potassium carbonate (K_2CO_3), hollow fiber membrane contactor (HFMC), modeling

1. Introduction

Carbon dioxide (CO_2) is widely recognized as a major contributor to global climate change. In recent years, the concentration of CO_2 in the atmosphere has increased significantly due to the intensive combustion of fossil fuels. Rising CO_2 emissions are a major cause of catastrophic environmental change that has led to growing interest in successful CO_2 capture [1]. Over the past decades, various technologies have been used for CO_2 capture. Chemical absorption by absorbents in trays and packed columns is the traditional method [2]. However, this method has economic and operational problems. Membrane contactor absorption is a new technology with many advantages, including: B. Prevention of interphase dispersion, high specific surface area, and compact size of the contactor [3]. In porous membrane contactors, absorption

 IntechOpen

typically occurs when CO_2 diffuses from the shell side through the membrane pores and contacts the liquid phase within the fibers. Therefore, there are no flooding, bubbling, channeling or entrainment problems associated with traditional absorber towers. Although membranes have been used commercially for gas separation since his 1980s, researchers have studied membranes over his 150 years [4–6]. Much research has been done on the development of membrane contactor systems and the mass transfer rates of membranes [7–10]. Membrane gas absorption has been introduced as one of the beneficial technologies to prevent CO_2 emissions due to its excellent mass transfer capabilities [11]. Many researchers compared his performance of CO_2 capture from flue gas using amine absorption (packed column or bubble column) as a conventional technique with membrane gas absorption technology [12, 13]. Their investigation showed that the membrane possesses a high surface-to-volume ratio, flexible operating characteristics, linear scaling, compact size, and modularity [3, 14, 15]. In addition, traditional absorber towers have many drawbacks such as channeling, foaming, flooding, aeration, and high operating and capital costs [16]. The use of membranes solves the above problem because CO_2 absorption occurs in the membrane contactor when the gas stream contacts the liquid phase on the other side of the membrane. In fact, the membrane acts as a barrier between the gas and liquid phases, preventing interpenetration of the gas and liquid phases. Additionally, increasing the membrane length may improve mass transfer between phases [16]. Furthermore, the membrane process has been introduced as an environmentally friendly alternative as a small amount of solvent, always lost to the atmosphere during the process [17]. Also, the liquid flow does not depend on the gas flow rate. However, the membrane process has some problems. The main problem is membrane wetting [18]. The most important criteria for choosing an absorbent in membrane gas absorption are the surface tension of the absorbent and its compatibility with the membrane material. Absorbents with low surface tension wet the pores of the membrane. Alkanolamine solvents such as MEA are the most commonly used absorbents for CO_2 capture due to their fast reaction rate with CO_2. However, amine solvents present challenges in terms of high regeneration energy and evaporative losses [19, 20]. Many researchers have spent a lot of time finding alternative absorbents to alkanolamines. A potential candidate is aqueous potassium carbonate (K_2CO_3). Potassium carbonate is an economical and environmentally friendly absorbent with low cost and low renewable energy for CO_2 absorption [21]. However, it slows down its reaction rate with CO_2, especially at low temperatures and low CO_2 partial pressures [22]. Adding an accelerator to potassium carbonate could be a viable approach to solve this problem [23].

2. Material and methods

In this work, the study was made by taking a volume element of the CO_2 absorption model using a membrane. We have a counter current flow from which the gas containing the CO_2 molecules enters the shell side ($Z = 0$); afterwards it crosses the membrane through the pores arriving in the tube side which contains the counter current solvent ($Z = l$). Also, an unsteady-state of a one-dimensional mathematical model has been considered and solved using ***pdepe*** MATLAB software. The model is developed for an element volume as shown in **Figure 1** and **Table 1**.

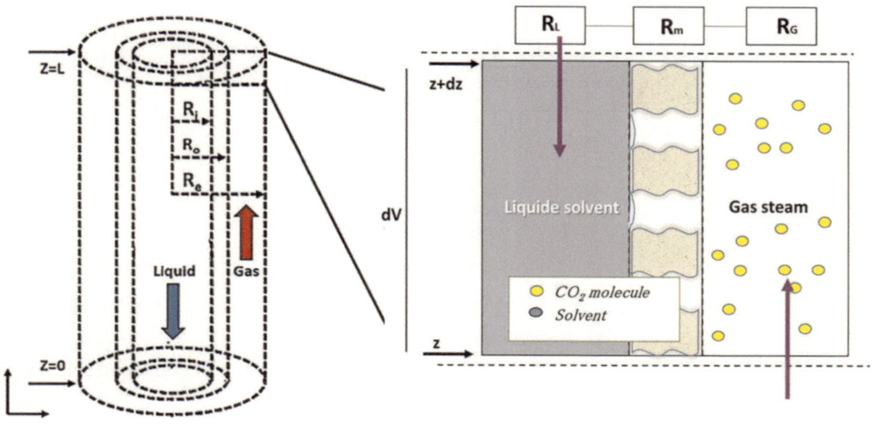

Figure 1.
Diagram of volume element mass transfer resistances.

Membrane material (abbreviation)	Reference	Gases separated	Solvent(s) trialed
Polypropylene (PP)	[25]	N_2/CO_2	MEA, DEA, MDEA, MEA/MDEA, DEA/2-amino2-methyl-1-propanol (AMP), a
	[26]	CO_2, CO, H_2, N_2, CH_4	Monoethanolamine, potassium carbonate
	[27]	CO_2	MEA
	[28]	N_2/CO_2	Monoethanolamine (MEA), [3 M]
	[29]	$N_2/CO_2/O_2$	Methyldiethanolamine (MDEA) [0.5e3 M]
Polyvinylidene fluoride (PVDF)	[30]	N_2/CO_2	Water
	[31]	N_2/CO_2	Diethanolamine (DEA)
Polytetrafuloroethylene (PTFE)	[26]	CO_2, CO, H_2, N_2, CH_4	Monoethanolamine [30 wt%], potassium carbonate [30 wt%]
	[32]	25% CO_2, 75% N_2	Monoethanolamine, MEA, 5 wt%
	[32]	CO_2, N_2	2-Amino-2-methyl-1-propanol (AMP)

Table 1.
Gas absorption membranes [24].

2.1 Mass transfer resistance in series model

The resistances in series in the hollow fiber membrane module **Figure 1**, can be written as following Eq. (1):

$$R_T = R_g + R_m + R_l \tag{1}$$

Where R_T is the total mass resistance, R_l is the mass resistance of the liquid phase, R_m is the mass resistance of HFMC and Rg is the mass resistance in the gas phase, as the resistance on the liquid side has been neglected. Then, by mass transfer term, "for example, see [33]":

$$\frac{1}{K_G} = \frac{d_{ext}}{d_{int}}\frac{1}{k_g} + \frac{d_{ext}}{d_{ln}}\frac{1}{k_m} + \frac{\mathcal{H}}{k_l \times Ha} \quad (2)$$

$$d_{ln} = \frac{d_{ext} - d_{int}}{ln\left(\frac{d_{ext}}{d_{int}}\right)} \quad (3)$$

$$Ha = \frac{\sqrt{Koh \times C_{k2co3} \times D_{co2,l}}}{k_l} \quad (4)$$

Where d_{int}, d_{ext} and Ha are the internal, external diameter and hatta number, respectively, k_g is the gas-phase mass transfer coefficient that is estimated using next following Eqs. (5)–(8), k_l is the liquid-phase mass transfer coefficient and k_m is membrane mass transfer coefficient [34].

2.1.1 *Gas-side mass transfer coefficient k_g*

$$Sh_g = \frac{k_g d_h}{D_{CO2,g}} = 0.6655 R_e^{0.5} SC^{0.33} \quad (5)$$

$$D_{CO2,g} = \frac{1.013 \times 10^{-2} T_g^{1.5}\left(\frac{1}{M_{CO2}} + \frac{1}{M_i}\right)^{0.5}}{P_g\left[\left(\sum_{CO2} v\right)^{\frac{1}{3}} + \left(\sum_i v\right)^{\frac{1}{3}}\right]^2} \quad (6)$$

$$R_e = \frac{\rho.d.v}{\mu} \quad (7)$$

$$Sc = \frac{\mu}{\rho.D_{CO2,g}} \quad (8)$$

2.1.2 *Membrane-side mass transfer coefficient k_m*

$$k_m = \frac{\varepsilon.D_{CO2,m}}{\tau.\delta} \quad (9)$$

Where ε and τ are the porosity and the tortuosity of the membrane, respectively. D_e represents the effective gas diffusivity, and δ represents thickness of the membrane [35].

$$\tau = \frac{(2 - \varepsilon)^2}{\varepsilon} \quad (10)$$

$$D_{CO2,m} = \frac{\varepsilon.D_{CO2,g}}{\tau} \quad (11)$$

2.1.3 *Liquid-side mass transfer coefficient k_l [35]*

$$Sh = \frac{k_l d_h}{D_{CO2,l}} = 0.552 R_e^{0.5} SC^{0.5} \quad (12)$$

$$D_{CO2,l} = 2.35 \times 10^{-6} \times e^{\left(-\frac{2119}{T}\right)}\left(\frac{\mu_{H2O}}{\mu_{K2CO3}}\right)^{0.8} \quad (13)$$

2.2 Mathematical model of a membrane contactor for a carbonated solvent

The reaction system of CO_2 with K_2CO_3 is given by [36, 37]:

$$CO_2 + 2KOH \rightleftarrows K_2CO_3 + H_2O \tag{14}$$

$$K_2CO_3 + H_2O + CO_2 \rightleftarrows 2KHCO_3 \tag{15}$$

$$KHCO_3 \rightleftarrows K^+ + HCO_3^- \tag{16}$$

The expression for the chemical kinetics of this reaction is given by the following equation [36]:

$$r_{(CO_2-K_2CO_3)} = K_{OH}.[OH^-][CO_2] \tag{17}$$

$$ln\,(K_{OH}) = 26.437 - \frac{5111.2}{T} \tag{18}$$

The general term for mass conservation is given as follows:

Accumulation = input − output + appearance–disappearance.

The following assumptions are used for the process modeling:

- Unidirectional (1D)
- Unsteady
- No wetting for membrane's pores

2.2.1 Gas phase

For the gas phase

$$\frac{\partial C_{CO2,g}}{\partial t} = D_{CO2,g}\frac{\partial^2 C_{CO2,g}}{\partial z^2} - V_g\frac{\partial C_{CO2,g}}{\partial z} - K_G a(\mathcal{H} \times C_{CO2,g} - C_{CO2,l}) \tag{19}$$

Where $C_{CO2,g}$ is the concentration of carbon dioxide in the shell side (gas), $C_{CO2,l}$ is the concentration of carbon dioxide in the tube side (liquid), Vg is the velocity on the gas phase, $D_{CO2,g}$ is the gas-side CO_2 diffusion coefficient (m^2/s), K_g is the overall transfer coefficient gas side (m/s), and $\mathcal{H}$ is henry constant.

2.2.2 Liquid phase

For CO_2

$$\frac{\partial C_{CO2,l}}{\partial t} = D_{CO2,l}\frac{\partial^2 C_{CO2,l}}{\partial z^2} - V_l\frac{\partial C_{CO2,l}}{\partial z} + K_G a(\mathcal{H} \times C_{CO2,g} - C_{CO2,l}) - K_{K2CO3} \times C_{CO2,l} \times C_{K2CO3} \tag{20}$$

t = 0 (s)	Position	Shell side	Tube side
$C_{CO_2,g}(z,0)=0$ $C_{K_2CO_3}(z,0)=40\ \text{mol/m}^3$	Z = 0	$C_{CO_2,g}(0,t)=15.3 mol/m^3$	$\frac{\partial C_{K_2CO_3}}{\partial z}(0,t)=0$
	Z = L	$\frac{\partial C_{CO_2,g}}{\partial z}(L,t)=0$	$C_{K_2CO_3}(L,t)=40\ \text{mol/m}^3$

Table 2.
Boundary and initial conditions of governing equations.

For K_2CO_3

$$\frac{\partial C_{K2CO3}}{\partial t}=D_{K2CO3,l}\frac{\partial^2 C_{K2CO3}}{\partial z^2}-V_l\frac{\partial C_{K2CO3}}{\partial z}-K_{OH}\times C_{CO2,l}\times C_{K2CO3} \tag{21}$$

See (**Table 2**).

3. Results and discussions

In this part the results show the simulation of the CO_2 capture process in a membrane contactor for the counter-current cases using the ***pdepe*** function of MATLAB, with this in mind, we will look at the effect of different parameters like solvent concentration, gas velocity and liquid velocity on CO_2 removal efficiency. The domain used in this study is shown in **Figure 1**. The parameters of the membrane used for the simulation are summarized in **Table 3**.

3.1 The effect of gas velocity

Figure 2 shows that as gas velocity increases, elimination efficiency decreases. This is because the contact time between the gas phase and the liquid phase decreases.

Parameters	Value	Unit
Material	PVDF	—
Membrane length	0.21	[m]
Membrane thickness (δ)	0.00034	[m]
Porosity (ε)	0.4585	
Nanoparticle morphology	—	Tubular
D_{CO_2},g	$1.39\times10^{-5}\ m^2\ s^{-1}$	$[m^2/s]$
D_{CO_2},m	1.58×10^{-5}	$[m^2/s]$
D_{CO_2},l	1.45×10^{-9}	$[m^2/s]$

Table 3.
Membrane's parameters [38].

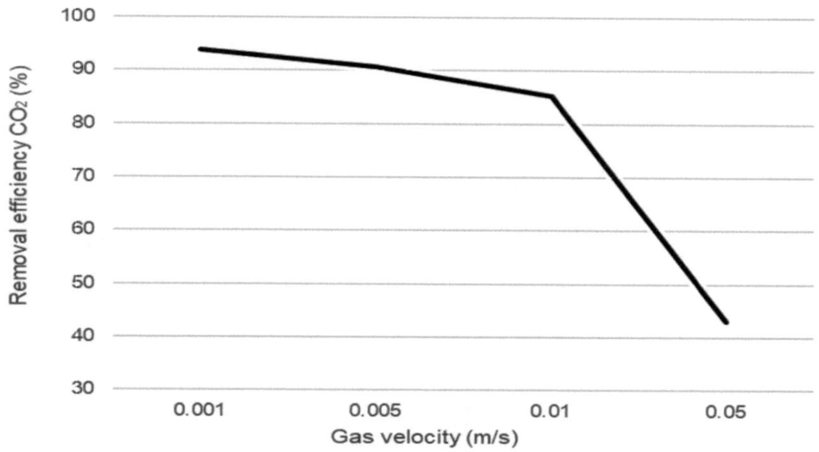

Concentration of K_2CO_3= 40(mol/m^3), liquid velocity =0.01(m/s),Porosity=0.4585, membrane length =0.21(m), t=100(s)

Figure 2.
Effect of gas velocity on CO_2 removal.

3.2 The effect of liquid velocity

Figure 3 shows that as the velocity of the liquid increases, the efficiency of CO_2 elimination increases. This is due to the increase in the speed of elimination of the CO_2 molecules that enter the liquid, which leads to an increase in the flow of CO_2 transferred.

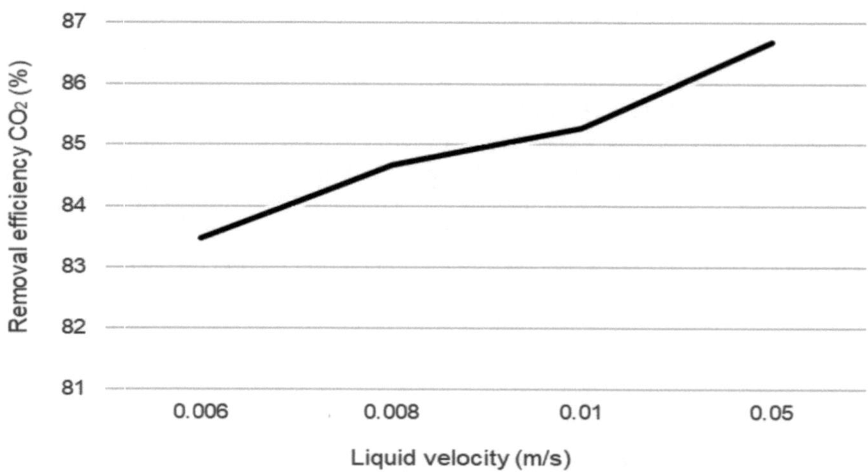

Concentration of K_2CO_3= 40(mol/m^3), gas velocity =0.01(m/s),Porosity=0.4585, membrane length =0.21(m), t=100(s)

Figure 3.
Effect of liquid velocity on CO_2 removal.

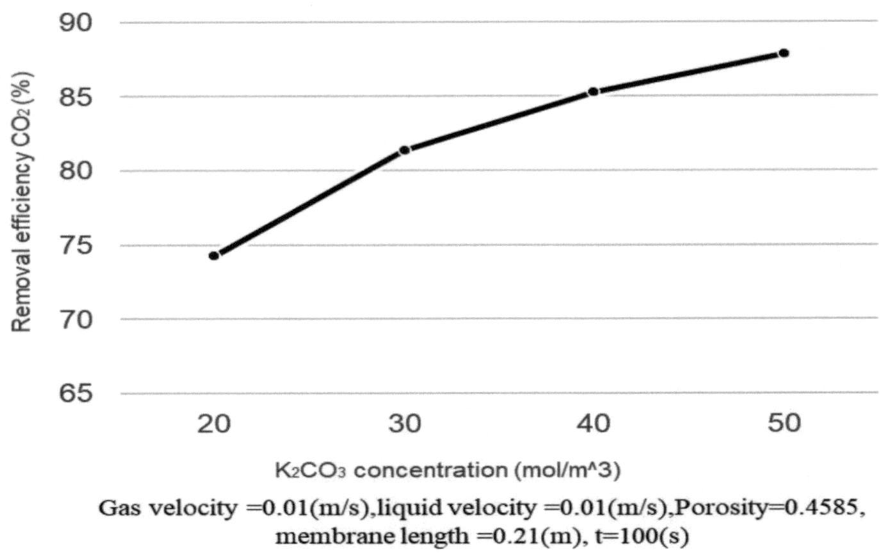

Figure 4.
Effect of loading concentration of K2CO3 on CO_2 removal.

3.3 The effect of inlet solvent concentration

It can be seen in **Figure 4**, that the efficiency of CO_2 elimination increases with increasing solvent concentration of the solvent. This is due to the increase in CO_2 reaction kinetics.

4. Conclusion

In a nutshell, robust and reliable mechanistic model and simulation methodology was validated and implemented to study the effects of concentration, gas velocity and liquid velocity on CO_2 capture by K_2CO_3 using membrane process to study the performance of the hollow fiber membrane contactor in terms of CO_2 removal. The CO_2 removal was increased by 52% in the range of 20–50 (mol/m^3) of Concentration of flow rate, also the results show that 13% increase (74–87%) on CO_2 capture while the increase of solvent concentration from 20 to 50 (mol/m^3), also for gas and liquid velocity from 0.001 to 0.05 and 0.006 to 0.05 (m/s) we have 52% and 3% increase of CO_2 capture respectively.

Nomenclature

a	effective interface area [m$^{2/}$m^3]
C_0	inlet concentration of CO_2 in shell side [mol/m^3]
C_{CO_2}	concentration of CO_2 [mol/m^3]
C_{K2CO3}	concentration of K_2CO_3 [mol/m^3]
C_{in}	concentration of K_2CO_3 in tube side [mol/m^3]
D_h	hydraulic diameter [m]
$D_{CO2,g}$	diffusion coefficient of CO_2 in the shell side [m^2/s]

D_{CO2},m	diffusion coefficient of CO_2 in membrane [m^2/s]
D_{CO2},l	diffusion coefficient of CO_2 in the tube side [m^2/s]
K_m	membrane mass transfer coefficient [m/s]
kg	gas-phase mass transfer coefficient [m/s]
kl	liquid-phase mass transfer coefficient [m/s]
K_G	gas-phase global mass transfer coefficient [m/s]
Re	Reynolds number
Sc	Schmidt number
Sh	Sherwood number
t	time [s]
T	temperature [k]
V_g	gas velocity [m/s]
V_l	liquid velocity [m/s]
μ	dynamics viscosity [kg/m.s]
ρ	density [kg/m^3]
$\mu H2O$	water viscosity [kg/m^3]

Author details

Mohamed Nadir Khelifi*, Ouacil Saouli, Anis Bouzeraib and Khaled Basta
Laboratory of Process Engineering for Sustainable Development and Health Products, Process Engineering Department, National Polytechnic School of Constantine, Algeria

*Address all correspondence to: nadir.khelifi@doctorant1.enp-constantine.dz

References

[1] Hoeven MD. CO2 Emissions from Fuel Combustion IEA Statistics. França: International Energy Agency; 2015

[2] Rao AB, Rubin ES. A technical, economic, and environmental assessment of amine-based CO_2 capture technology for power plant greenhouse gas control. Environmental Science & Technology. 2002;**36**:4467-4475

[3] Gabelman A, Hwang S-T. Hollow fiber membrane contactors. Journal of Membrane Science. 1999;**159**:61-106

[4] Graham T. Notice of the singular inflation of a bladder. Journal of Membrane Science. 1995;**100**:9

[5] Graham T. On the law of the diffusion of gases. Journal of Membrane Science. 1995;**100**:17-21

[6] Baker RW. Future directions of membrane gas separation technology. Industrial & Engineering Chemistry Research. 2002;**41**:1393-1411

[7] Bhide BD, Voskericyan A, Stern SA. Hybrid processes for the removal of acid gases from natural gas. Journal of Membrane Science. 1998;**140**:27-49

[8] Powell CE, Qiao GG. Polymeric CO_2/N_2 gas separation membranes for the capture of carbon dioxide from power plant flue gases. Journal of Membrane Science. 2006;**279**:1-49

[9] Favre E. Carbon dioxide recovery from post-combustion processes: Can gas permeation membranes compete with absorption. Journal of Membrane Science. 2007;**294**:50-59

[10] Pires JCM, Martins FG, Alvim-Ferraz MCM, Simões M. Recent developments on carbon capture and storage: An overview. Chemical Engineering Research and Design. 2011;**89**:1446-1460

[11] Li J-L, Chen B-H. Review of CO2 absorption using chemical solvents in hollow fiber membrane contactors. Separation and Purification Technology. 2005;**41**:109-122

[12] Bounaceur R, Lape N, Roizard D, Vallieres C, Favre E. Membrane processes for post-combustion carbon dioxide capture: A parametric study. Energy. 2006;**31**:2556-2570

[13] Okabe K, Mano H, Fujioka Y. Separation and recovery of carbon dioxide by a membrane flash process. International Journal of Greenhouse Gas Control. 2008;**2**:485-491

[14] Rangwala HA. Absorption of carbon dioxide into aqueous solutions using hollow fiber membrane contactors. Journal of Membrane Science. 1996;**112**:229-240

[15] Klaassen R, Feron P, Jansen A. Membrane contactors in industrial applications. Chemical Engineering Research and Design. 2005;**83**:234-246

[16] Eslami S, Mousavi SM, Danesh S, Banazadeh H. Modeling and simulation of CO2 removal from power plant flue gas by PG solution in a hollow fiber membrane contactor. Advances in Engineering Software. 2011;**42**:612-620

[17] Peters L, Hussain A, Follmann M, Melin T, Hägg MB. CO_2 removal from natural gas by employing amine absorption and membrane technology—A technical and economical analysis. Chemical Engineering Journal. 2011;**172**:952-960

[18] Wang R, Zhang HY, Feron PHM, Liang DT. Influence of membrane wetting on CO_2 capture in microporous hollow fiber membrane contactors. Separation and Purification Technology. 2005;**46**:33-40

[19] Thee H, Nicholas NJ, Smith KH, da Silva G, Kentish SE, Stevens GW. A kinetic study of CO_2 capture with potassium carbonate solutions promoted with various amino acids: glycine, sarcosine and proline. International Journal of Greenhouse Gas Control. 2014;**20**:212-222

[20] Zare P, Keshavarz P, Mowla D. Membrane absorption coupling process for CO_2 capture: Application of water-based ZnO, TiO_2, and multi-walled carbon nanotube nanofluids. Energy & Fuels. 2019;**33**:1392-1403

[21] Hu G, Smith KH, Wu Y, Kentish SE, Stevens GW. Screening amino acid salts as rate promoters in potassium carbonate solvent for carbon dioxide absorption. Energy & Fuels. 2017;**31**:4280-4286

[22] Hu G, Nicholas NJ, Smith KH, Mumford KA, Kentish SE, Stevens GW. Carbon dioxide absorption into promoted potassium carbonate solutions: A review. International Journal of Greenhouse Gas Control. 2016;**53**:28-40

[23] Izaddoust A, Keshavarz P. Experimental and theoretical study of CO_2 absorption with piperazine-promoted potassium carbonate solution in hollow Fiber membrane contactors. Energy & Fuels. 2017;**31**:9790-9799

[24] Rajab K, Kathryn M, Haibo Z, et al. Membrane-based carbon capture from flue gas: A review. Journal of Cleaner Production. 2015;**103**:286-300

[25] Wang Z, Fang MX, Yan SP, Yu H, Wei CC, Luo ZY. Optimization of blended amines for CO_2 absorption in a hollow-fiber membrane contactor. Industrial and Engineering Chemistry Research. 2013;**52**:12170-12182

[26] Scholes CA, Simioni M, Qader A, Stevens GW, Kentish SE. Membrane gasesolvent contactor trials of CO_2 absorption from syngas. Chemical Engineer. 2012;**195–196**:188-197

[27] Khaisri S, Demontigny D, Tontiwachwuthikul P, Jiraratananon R. Comparing membrane resistance and absorption performance of three different membranes in a gas absorption membrane contactor. Separation and Purification Technology. 2009;**65**: 290-297

[28] Bottino A, Capannelli G, Comite A, Di Felice R, Firpo R. CO2 removal from a gas stream by membrane contactor. Separation and Purification Technology. 2008;**59**:85-90

[29] Yan S-P, Fang M-X, Zhang W-F, Wang S-Y, Xu Z-K, Luo Z-Y, et al. Experimental study on the separation of CO_2 from flue gas using hollow fiber membrane contactors without wetting. Fuel Processing Technology. 2007;**88**: 501-511

[30] Zhang H-Y, Wang R, Liang DT, Tay JH. Theoretical and experimental studies of membrane wetting in the membrane gaseliquid contacting process for CO_2 absorption. Journal of Membrane Science. 2008;**308**:162-170

[31] Mansourizadeh A, Ismail AF. Hollow fiber gaseliquid membrane contactors for acid gas capture: A review. Journal of Hazardous Materials. 2009;**171**:38-53

[32] Yeon SH, Sea B, Park YI, Lee KH. Determination of mass transfer rates in PVDF and PTFE hollow fiber membranes for CO_2 absorption.

Separation Science and Technology. 2003;**38**:271-293

[33] Kim Y-S, Yang S-M. Absorption of carbon dioxide through hollow fiber membranes using various aqueous absorbents. Separation Purification Technology. 2000;**2000**:21

[34] Luo X, Hartono A. Comparative kinetics of carbon dioxide absorption in unloaded aqueous monoethanolamine solutions using wetted wall and string of discs columns. Chemical Engineering Science. 2012;**82**:31-43

[35] Chabanon E, Roizard D. Modeling strategies of membrane contactors for post-combustion carbon capture: A critical comparative study. Chemical Engineering Science. 2013;**87**:393-407

[36] Tang X. Thermodynamic Properties of Acid Gases in Mixture with Natural Gas and Water. 2011

[37] Álvaro P-SK, Eckehard M, Bernd R, et al. Solubility of CO_2 in aqueous solutions of KCl and in aqueous solutions of K_2CO_3. Journal of Chemical & Engineering Data. 2007;**52**(3):817-832

[38] Yan C, Zia R, Nayef G, et al. Intensification of CO_2 absorption using MDEA-based nanofluid in a hollow fibre membrane contactor. Scientific Reports. 2021;**11**(1):2649

Section 2

New Insights on Distillation

Chapter 5

A Systematic Review of the Literature on Steady-State Reactive Distillation Modeling and Simulation: Challenges and Opportunities

Vilmar Steffen, Maiquiel Schmidt de Oliveira and Edson Antonio da Silva

Abstract

Reactive distillation has already been extensively studied, many applications have been evaluated by simulation and laboratory experiments, and many modifications in the configuration of the column are being proposed, such as the reactive dividing wall column. The steady state plays a key role because it is required to operate a process steadily at the optimal point, and it is necessary to use control strategies to keep the operation at this point. Thus, this chapter presents a systematic literature review, about the modeling and simulation of reactive distillation in steady state presenting some bibliometric results. It also presented some content related to the variations in the configurations of a reactive distillation column, mathematical modeling of the process, and computational simulation. Lately, simulation studies reported in the literature about this subject have mainly used sequential modular simulators like Aspen Plus. But, the solution process of the model's equations still can be improved. Some opportunities are pointed out, like the use of heuristic algorithms for the solutions of the mathematical equations, better initial guess algorithm for the unknowns, and the use of a more generic scheme for representation of the mathematical model of a reactive distillation stage.

Keywords: reactive distillation, modeling, simulation, initial guess, steady-state, tearing equations, inside-out, numerical solution, *Methodi Ordinatio*

1. Introduction

Many traditional processes related to the chemical industries comprise different stages, in which reaction and separation are among the most important ones [1]. The reaction occurs in different reactor formats (tank or tubular) and operation types (batch or continuous), and the reaction can need some catalyst (homogeneous or heterogeneous). Regarding the separation step, distillation is one of the most widely used unit operations to separate mixtures in chemical industry applications [2, 3]. In recent years, there has been permanently increasing interest in the development of hybrid processes combining reaction and separation, among which can be highlighted reactive distillation processes (configuration in which reaction and separation happen in the same column) [1, 4–9], known as a reactive distillation column (RDC), and the phenomenon is, thereby, referred to as reactive distillation (RD) process [10–16]. In this way, several new processing methods have been developed and commercialized [17]. When reactive distillation is compared to conventional processes, one can see that reactive distillation results in a simpler flowchart with fewer recycle streams and a lower number of separation units [18].

Distillation is considered as a mature technology and has been widely used in many processes over the years; however, some improvements are being made in order to reduce energy consumption, since it is well known for its high energy requirements, and improve thermodynamic efficiency [19–21]. Furthermore, world energy consumption is increasing at a steady rate for many reasons like population growth, industrialization and transportation [22]. Process intensification is a promising tool to increase the sustainability of chemical processes by using novel apparatuses and techniques [3, 23].

Reactive separation conveniently combines the production and removal of one or more products, enabling enhancement of conversion, simplicity, selectivity and yield, which can present various technical, environmental and economic benefits compared to reaction and separation taking place separately [4, 8, 22, 24–35].

Another advantage of some reactive distillation applications is that the reaction temperature is easy to control in a reactive distillation column. The temperature of each stage is always the bubble point, and it depends on the reactive mixture and system pressure. Therefore, the reaction temperature can be controlled by adjusting the system pressure [36]. The heat of an exothermic reaction is used for the vaporization of liquids, reducing the reboiler duties. The higher temperature value in the reaction section is limited by the boiling point of the mixture of the components involved in the process; thus, the formation of hot spots is severely reduced [10, 37, 38].

The thermodynamic limitations, for example, the presence of azeotropes, make conventional separation systems complex and expensive but can be overcome using reactive distillation. Of course, the reaction temperature must be suitable to the column conditions [24, 39], and the reactants and products relative volatilities must be such that products are removed and reactants retained inside the column [40].

Reversible reactions can be limited to equilibrium, which is difficult to overcome because they have reached a point in which higher product purity is not possible if the reaction conditions are kept [16]. Reactive distillation can be used to overcome this problem by continuous removal of products from the reaction zone [22, 37]. If a liquid-phase reaction must be carried out with a large excess of one reactant, reactive distillation is potentially attractive because it can be carried out closer to stoichiometric feed conditions, avoiding recycling costs [41]. Side reactions can be dampened or even avoided by the constant separation of the products. Also, the formation of azeotropes can be prevented [10, 42].

Reactive distillation has been employed in industry for many decades [24]. However, the potential of reactive distillation has not yet been fully tapped, and there is still ongoing research to improve [43]. Reactive distillation can be used with a wide variety of reactions; among suitable reactive distillation processes, already studied, are acetylation, alkylation, amination, dehydration, etherifications, nitrations, esterifications, hydrolysis, isomerization, transesterifications, polycondensations, and halogenations [1, 7, 44], offering an attractive alternative for the production of many important industrial chemicals [45]. With respect to the exploration of the potential of the reactive distillation process, the steady-state simulation of a reactive distillation column is an important step in the implementation of the technology and the discovering of its potential advantages, and it is essential for design, investigation of new applications, control and optimization [46, 47].

A literature review is fundamental for knowing the cutting edge of a scientific subject and also can be used to find some interesting questions that have not yet been answered, but there is so much scientific content that is important to systematize the review process, because in a review not systematize the research tends to cite and comment the publications that he/she like or known most, and the systematic literature review can help to avoid this making the review process reproducible. Also, a systematic literature review can provide some bibliometric data that can be used to see where (countries, institutions, etc.) the subject is studied most [48, 49]. Among the many systematic literature review methods proposed in the literature, like PRISMA [50], ProKnow-C [51], SIMILAR [52] and NIRP [48], among others, there are some that help the researcher to obtain a rank of the importance of the papers in the final portfolio, among which is *Methodi Ordinatio* [53, 54]. Knowing the importance of many papers is important to deciding which must receive more and less attention.

In this work, a systematic review of the literature (applying the *Methodi Ordinatio*) was carried out about the modeling and simulation of steady-state (SS) reactive distillation column aiming to present a collection of the most used methods and tools used in this research area and find some future studies that should be interesting to consider. The steps of the *Methodi Ordination* are presented in Section 2. Also, in this section, the choices made in the application of the method are presented, as well as some data collected in this application. After getting the list of articles, some aspects related to the subject were collected and presented in Sections 3, 4, and 5. Some challenges and opportunities are presented in six, and the conclusions about the review are outlined in Section 7.

2. Methodology

This section presents the *Methodi Ordination*, the data obtained in each step of this method, and some bibliometric results.

2.1 *Methodi Ordinatio*

In this research, a systematic review of the literature was carried out, aiming at publications involving the steady-state modeling and simulation of reactive distillation. The *Methodi Ordinatio*, proposed by [53, 54], was used in the construction of the literature review to determine which is the most relevant research in the field.

Literature reviews are important in order to organize papers and to find possible lacks and opportunities in the research field. The *Methodi Ordinatio* is a systematic literature review method based on an index (*InOrdinatio*) to generate a ranking grouping of the most important papers. The nine steps of the method are presented below:

1. *Research topic definition*: before you start research, you must define what you want to look for;

2. *Preliminary search*: some tests can be carried out in order to clarify what you are searching;

3. *Definition of keywords, databases and time delimitation*: the previous step can help you to define the combination of keywords. Beyond that, you should choose the databases more relevant to the research topic and, in some cases, it is interesting to limit the range of time for the review;

4. *Definitive search in the database*: Finally, you have the necessary data to do the research in the databases. In this step, and in some of the upcoming ones, it is primordial to use a reference manager software (like EndNote, Mendeley, JabRef, Zotero, etc.);

5. *Filtering procedure*: mainly if the search is carried out in more than one database, it is normal to have some duplicate papers, and these must be resolved. Some results present no data, which results in the deletion. Also, it is accepted only for research papers. Conference papers, books, book chapters, and review papers are not accepted, even if it is a good research item, because the impact factor indicator is not available for this kind of research item, which makes it impossible to calculate the index (*InOrdinatio*) which is used to generate the rank of importance of the research items;

6. *Searching for papers data*: some data are necessary for each paper is necessary for the *ordinatio* index calculation that is carried out in the next step;

7. *InOrdinatio*: the order of the article's importance is determined by applying the data collected in the last step in Eq. (1)

$$\text{InOrdinatio} = (\text{IF}/1000) + \alpha[10 - (\text{ResearchYear} - \text{PublishYear})] + \text{Ci} \tag{1}$$

   where IF is the journal impact factor in which the article was published, α is a value (from 1 to 10) related to the subject importance and newness, ResearchYear is the year of the data collection (2023 for this review), PublishYear is the year of the article publication, and Ci is the number of times the paper has already been cited.

8. *Finding texts in full format*: it is interesting to perform this action at this point because at this point the number of papers is much smaller;

9. *Systematic reading and analysis of articles*: some interesting results can be found by analyzing the bibliometric results and mainly by effectively reading the papers.

After all these steps, some lacks and possible opportunities can be found.

2.2 Review steps data

Each step of a systematic review of literature can result in some data that is not necessarily a result, but it is important to understand the procedure. Below it is presented some of them:

1. *Research topic definition*: the chosen subject was modeling and simulation of reactive distillation in steady state;

2. *Preliminary search*: this step was fundamental to build the keyword combination;

3. *Definition of keywords, databases and time delimitation*: Due to the subject, it was not set time delimitation; the databases chosen are ScienceDirect, Scopus and Web of Science and the keywords combination is:

   ("reactive distillation" OR "RD") AND "steady state" AND ("simulation" OR "modelling" OR "modeling")

4. *Definitive search in the database*: this step resulted in a total of 547 research items:

   - ScienceDirect: 135;
   - Scopus: 293;
   - Web of Science: 119.

5. *Filtering procedure*: some deletions of research items were made because of:

   - Duplicates: 174;
   - No research item Data: 8;
   - Type of research item (conference paper, review paper, book chapter, etc.): 68;
   - Out of scope (by reading title, keyword and abstract): 144.

   Resulting in 153 papers. Analyzing this step, one can see that title, keywords and abstract must be well written and properly represent the paper content; otherwise, the paper can be incorrectly excluded from the review.

6. *Searching for papers data*: All data of the final portfolio were found;

7. *InOrdinatio*: With all data collected in the previous step it was applied Eq. (1) for the ranking composition. The journal impact factor used was JCR. Due to the recentness of the subject is not so relevant in modeling and simulation, a small value of α must be chosen, we choose $\alpha = 1$ because there are publications about the subject over more than three decades, therefore it is not a recent research subject. This is subjective choice is a weakness of the *Methodi Ordinatio*; therefore, there are already some works proposing some improved methods for ranking papers [48, 49]. In this step, nine papers were removed because they resulted in a negative index. The number of citations is from Google Scholar [55, 56]. So, there are 144 articles in the final portfolio.

8. *Finding texts in full format*: All articles were downloaded.

9. *Systematic reading and analysis of articles*: Some important features can be found in this step, and some of them are presented in the next section.

2.3 Bibliometric results

After obtaining the final portfolio, some important information can be found by analyzing the article's bibliometric data. **Figure 1** shows the number of articles per

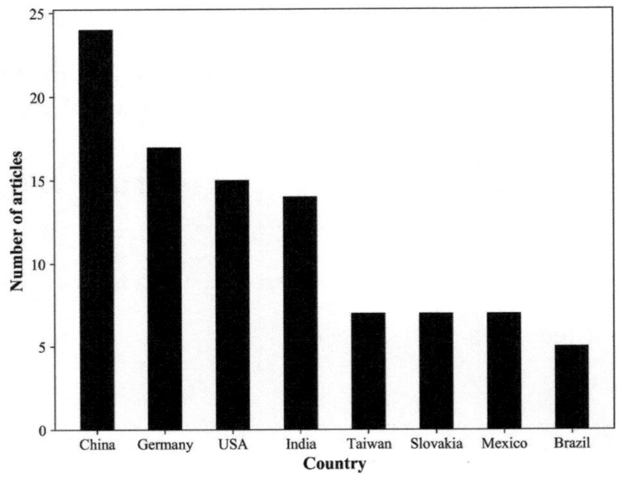

Figure 1.
Main countries of steady-state reactive distillation modeling and simulation articles.

country of the institution in which the first author was affiliated when the work was carried out. In this figure, one can see that just four countries (namely, China, Germany, the United States of America and India) are responsible for 70 articles, almost half of the portfolio. Beyond these countries presented in **Figure 1**, there are three other countries with three articles each, five countries with three articles each, seven countries with two papers each, and seven countries with one article each. In total, the portfolio contains the first authors affiliated with 30 different countries.

Figure 2 shows the journals with a higher number of articles in the final portfolio. Just five journals (namely, Industrial and Engineering Chemistry Research, Chemical Engineering and Processing: Process Intensification, Computers and Chemical Engineering, Chemical Engineering Science and Chemical Engineering Research and Design) are responsible for 79 articles, more than half of the portfolio. Beyond the 18 Journals presented in **Figure 2**, there are 19 other journals with one article each, that is, the articles from the final portfolio were published in 37 different journals.

As can be seen in **Figure 3**, the first publication of the portfolio took place in 1991; after this, there was no publication for 5 years, and then the number of publications

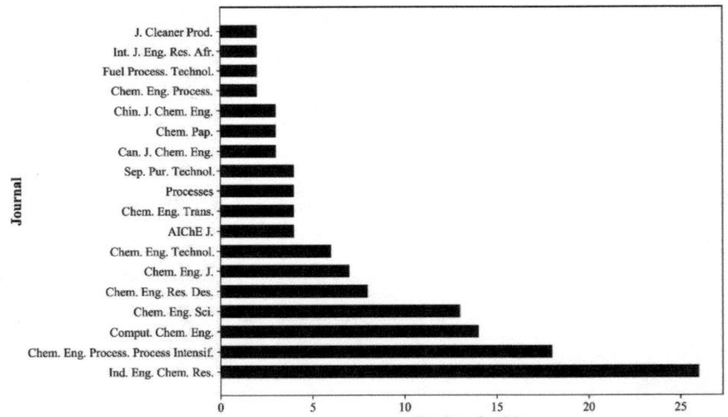

Figure 2.
Main journals of steady-state reactive distillation modeling and simulation articles.

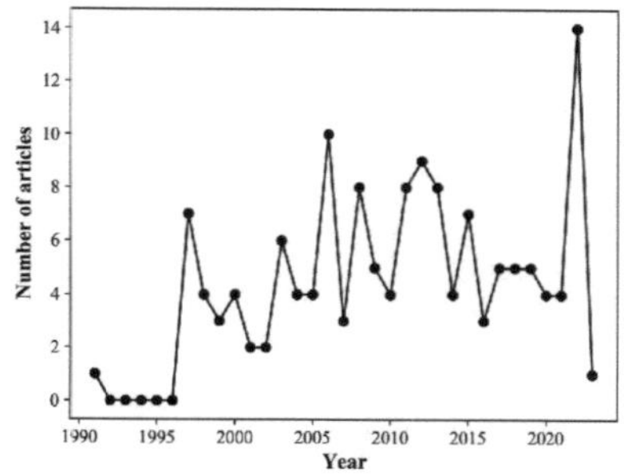

Figure 3.
Number of articles on steady-state reactive distillation modeling and simulation over the years.

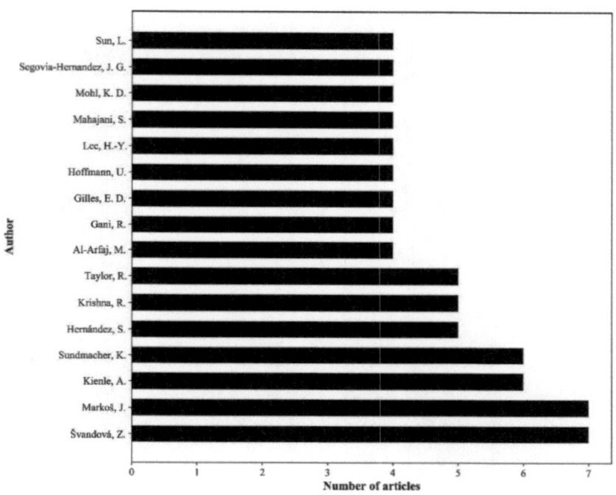

Figure 4.
Main authors of steady-state reactive distillation modeling and simulation articles.

has varied over the years, with a slight tendency to rise. Among all these years, 2022 has the greatest number of publications.

In the final portfolio, there are 351 different authors, among which 249 authors have only one paper, 66 authors have 2 papers from the final portfolio, 20 authors have 3 papers, 9 authors with 4 paper, 3 authors have 5 papers and only 2 authors have 6 and 7 papers from the final portfolio. **Figure 4** shows the main authors, in which Švandová, Z. and Markoš, J., the authors who have the most publications in the final portfolio, have seven papers because they published the same seven works; a similar situation occurs with Taylor, R. and Krishna, R., who published the same five articles.

3. Reactive distillation column configurations

Reactive flash is the simplest reactive distillation configuration where separation is carried out in only one vaporization stage [57]. The reactive distillation column is very similar to the conventional distillation column, as illustrated in **Figure 5a**, in which the separation process occurs through many stages. If the reaction has a rate high enough to be conducted in pressure and temperature of a distillation column conditions without the need for a catalyst, the reaction takes place in all stages of the column, even in the reboiler and condenser, but normally it is considered that the reaction does

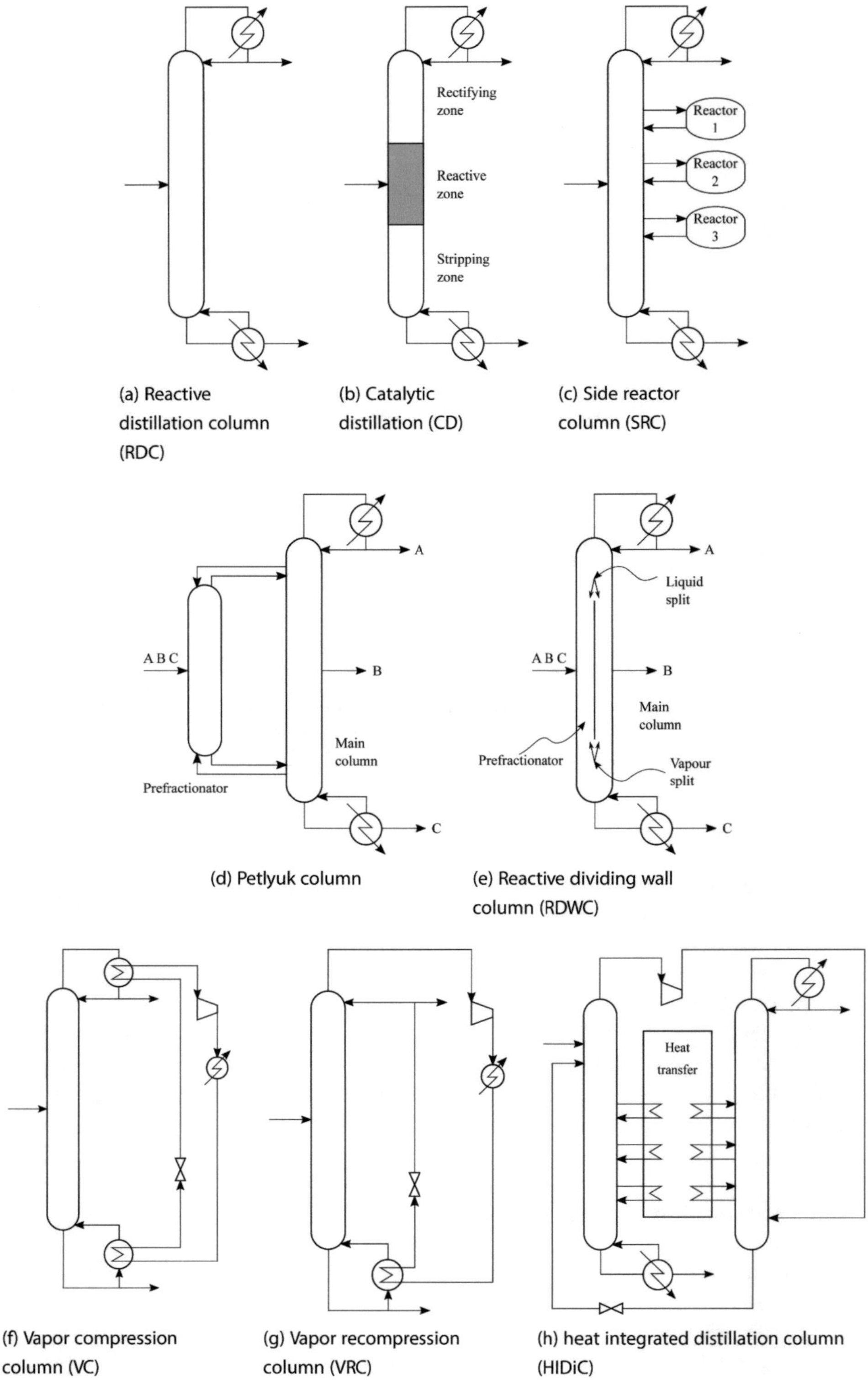

Figure 5.
Variations on reactive distillation.

not occur in the reboiler and condenser to simplify the solution of the mathematical model.

In the cases that a catalyst is applied in the reactive distillation, that is, a catalyst column, just some stages are reactive, so it is possible to divide the column into three zones like presented in **Figure 5b**, namely the rectifying zone, reactive zone and stripping zone. If the reaction is heterogeneously catalyzed, the reactive zone is formed by the stages packed with catalyst. In the case of a homogeneously catalyzed reaction, the reactive section is defined by the location of the feed stream and catalyst properties; for example, if the catalyst can be considered to be non-volatile, the reactive zone is formed by feed stage and the stages below that [58].

A side reactor column (SRC), like the one presented in **Figure 5c**, is a distillation column connected to one or more reactors that can be used if the distillation conditions do not match the conditions for good reaction rates, the column operates at favorable conditions for separation, while the reactor operates at reaction kinetics favorable conditions [59].

In some cases, the separation of mixtures taking place in a conventional distillation sequence can be thermally inefficient. In these cases, one can turn to a thermally coupled distillation sequences (TCDS) solution (that has been used for the separation of multicomponent mixtures achieving energy savings) like the Petlyuk column (total thermal coupled solution) [19], illustrated in **Figure 5d**, in which it is used a prefractionator column to separate, for example, a mixture of A, B and C, with A the lightest and C the heaviest components in the mixture. An even better solution that can save energy and capital costs is the use of the prefractionator and the main column in a single shell, as shown in **Figure 5e**. This solution is called a dividing wall column (DWC) because it is a single shell divided by a wall in the middle of some stages of a conventional distillation column. Dividing wall columns enables the separation of four pure fractions. If the dividing wall column is used for a reactive distillation process, it is called a reactive dividing wall column (RDWC).

Reactive dividing wall column (RDWC) is a process in which reactive distillation and dividing wall column are integrated, being that it simultaneously has the advantages of both processes [60]. For heterogeneous catalytic reactions, the reaction zone is normally present on the feed side of the column where the reaction rate is higher because it is where the reactant concentrations are higher, but there are some propositions in the literature for more than one reactive zone [22]. By combining an enzymatic catalyst and a reactive dividing wall column, we have an enzymatic catalyzed reactive dividing wall column (eRDWC).

Heat pump (HP) distillation is a technology that uses the latent heat from the vapor top product as the heat source in the reboiler [61]. There are two mainly conventional ways to use the heat pump in a distillation column (or reactive distillation column): the vapor compression column (VC) and the vapor recompression column (VRC). As shown in **Figure 5f** and **g**, in the VC case, the working fluid is evaporated in the condenser, compressed and heated over the reboiler temperature, condensed in the reboiler and cooled down, through a valve, below the condenser temperature, and in the VRC case the working fluid is the vapor leaving the top of the column, and after cooled down through the valve part of the working fluid is refluxed to the column. If a vapor compression or recompression is applied to a reactive distillation column, we have a heat pump for reactive distillation (HPRD).

To improve energy efficiency, the heat integration concept can be applied to a distillation column. Heat-integrated reactive distillation is beneficial for energy saving and economy compared to conventional reactive distillation [62]. The idea of heat integration is the use of hot process streams to raise the temperature of cold process streams. There are several heat integration techniques, one of which is called heat integrated distillation column (HIDiC) technique, **Figure 5h**. This technique is called heat-integrated reactive distillation column (HIRDiC) if applied to a reactive distillation process [63, 64].

Many other configurations can be found in the literature that are applied in the distillation and reactive distillation, like pressure swing distillation (PSD) or pressure swing reactive distillation (PSRD), which uses two columns operating at two different pressures applied to process separations of pressure-sensitive azeotropes [65].

Using circulating reflux (pump around), normally from the lower stages of a column to stages, makes a stream in a higher temperature to be recirculated in the column, increasing the driving force for heat transfer, which is normal, for example, in a crude oil distillation column that is a multi-product process with many side stream.

Reactive-extractive dividing wall column (REDWC) is a combination of reactive distillation, extractive distillation and dividing wall column, resulting in a highly integrated process that demands less physical space and has a strong interrelation among variables [66].

There are also some combinations that work very well for a specific process that normally need three or more columns, which makes it impractical to be listed here.

Although there are so many configurations in the reactive distillation columns, among which some were presented in this section, the modeling is very similar because it can be made stage-by-stage; therefore, the difference is due to the connections among the stages. Thus, the modeling presented in the next section is focused in the model of one generic stage.

4. Modeling

Although the steady state is not the reality in a chemical process, it is very important to understand the process. The steady-state can show us the limitations of the process, that is, the maximum the process can yield with the current configuration. The steady-state can also be used for optimization and consequently can be important in control because it is desired that the process operates all the time at the optimal point, and the use of control strategies to keep the operation at this point. Because of all these reasons, the mathematical model of a reactive distillation column operating in a steady state is very important to evaluate some features of the process.

A reliable mathematical model is fundamental for understanding and analyzing the behavior of a reactive distillation process [67]. The development of a phenomenological mathematical model of a process is based on applying conservation laws (mass, energy and momentum) and constitutive relations (like chemical equilibrium or chemical reaction kinetics, phase equilibrium, etc.). A mathematical model must represent (approximately) the real behavior of some relevant properties of a process and can be very important to analyze an existing plant or to evaluate the technical viability of a new chemical process plant [68]. Most of the reactive distillation models are originally adapted from conventional distillation calculations based on equilibrium stage model [69]. The reactive distillation column is modeled stage-by-stage with adequate thermodynamic behavior and chemical reactions [14].

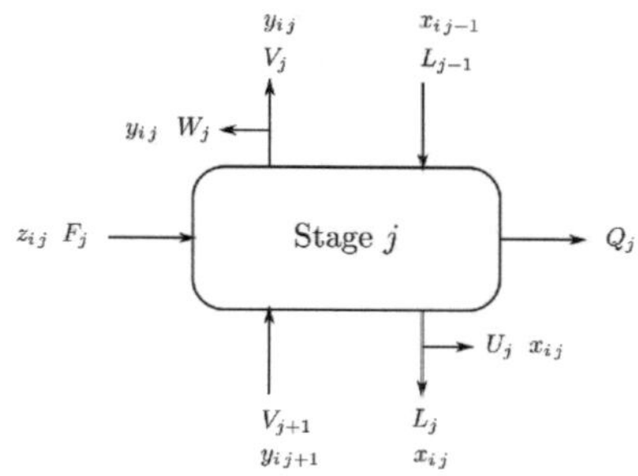

Figure 6.
Schematic of a generic stage in a reactive distillation column.

For the application of conservation law, for a distillation column is applied in each stage, it is fundamental to know the mass and energy flows. For example, **Figure 6** shows a schematic of the flows of a generic stage in a reactive distillation column. This generic stage can be used to represent any stage in the column, even the condenser (stage 1) and reboiler (stage n, last stage). For this reason, it is considered in all stages the liquid side flow from stage j (U_j), vapor side flow from stage j (W_j), heat load from stage j $\left(Q_j\right)$ and flow rate of the feed stream to stage j (F_j). Furthermore, each stage has an liquid inlet from stage right above (L_{j-1}), a vapor inlet from the stage right below (V_{j+1}) and liquid and vapor outlets from stage j (respectively, L_j and V_j). If it is considered a homogeneous stage, the liquid and vapor outlets from the stages have the same composition of the stage content, that is, the mole fraction of component i in stage j of liquid da vapor phases (x_{ij} and y_{ij}). The feed stream can be in liquid or vapor phase, so z_{ij} is the mole fraction of component i in the feed stream of stage j.

The equilibrium model of a stage in a reactive distillation column is represented by MESH equations [67, 70–74]:

- Material balance, global and per component, are represented, respectively, by Eqs. (2) and (3):

$$F_j + R_j + V_{j+1} + L_{j-1} - (L_j + U_j) - (V_j + W_j) = 0 \tag{2}$$

$$z_{ij}F_j + r_{ij} + y_{ij+1}V_{j+1} + x_{ij-1}L_{j-1} - x_{ij}(L_j + U_j) - y_{ij}(V_j + W_j) = 0 \tag{3}$$

where r_{ij} is the generation rate of component i on stage j and R_j is the total rate of generation on stage j. The relation between these variables is given by the sum of the generation of all m components in the process, presented by Eq. (4).

$$R_j = \sum_{i=1}^{m} r_{ij} \tag{4}$$

- Equilibrium equation. Normally, the liquid-vapor equilibrium, Eq. (5), is considered:

$$y_{ij} = K_{ij}x_{ij} \tag{5}$$

where K_{ij} is liquid-vapor equilibrium (EQ) relation for the component i on stage j, that for a gamma-phi formulation is given by Eq. (6):

$$K_{ij} = \frac{\gamma_{ij}\phi_i^{sat}P_i^{sat}}{\hat{\phi}_{ij}P} \exp\left(\int_{P_i^{sat}}^{P} \frac{V_i^L}{RT}\,dP\right) \tag{6}$$

where γ_{ij} is the activity coefficient of component i on stage j, $\hat{\phi}_{ij}$ is the fugacity coefficient of component i on stage j, ϕ_i^{sat} is the fugacity coefficient of pure component i in saturation state, P is the pressure, P^{sat} is the saturation pressure, R is the ideal gas constant, T is the temperature, and V_i^L is the component i volume in liquid phase.

- Summation equations. The sum of mole fractions, in the liquid and vapor phases, must be equal to one in all n stages, as presented in Eqs. (7) and (8):

$$\sum_{i=1}^{m} x_{ij} = 1 \tag{7}$$

$$\sum_{i=1}^{m} y_{ij} = 1 \tag{8}$$

- Heat (enthalpy) balance, that is, energy balance presented in Eq. (9):

$$h_j^F F_j + H_{j+1}V_{j+1} + h_{j-1}L_{j-1} - h_j\left(L_j + U_j\right) - H_j\left(V_j + W_j\right) - Q_j = 0 \tag{9}$$

where h_j is the enthalpy of liquid stream flow from stage j, h_j^F is the enthalpy of feed stream flow to stage j and H_j is the enthalpy of vapor stream flow from stage j. For this energy balance, the enthalpies must be calculated using the heat of formation as a reference state, so it is not necessary to use an additional term of heat of reaction [1, 75].

Normally, the degrees of freedom in an algebraic system of equations representing the behavior of physical properties in a distillation column are greater than zero. To work around this problem, it is possible to specify some unknown values, but it is habitual to use ratios like those presented in Eqs. (10) and (11):

$$U_j = \eta_j^L L_j \tag{10}$$

$$W_j = \eta_j^V V_j \tag{11}$$

where η_j^L is the ratio between the liquid side stream and the liquid stream outputting stage j and inputting the near stage $j+1$ and η_j^V is the ratio between the vapor side stream and the vapor stream outputting stage j and inputting the near stage $j-1$.

To evaluate the generation rate of each component in each stage, it is necessary to evaluate the extent of reaction of each reaction in each stage, as shown in Eq. (12):

$$r_{ij} = \delta_j \sum_{k=1}^{p} \nu_{ki} \xi_{kj} \tag{12}$$

where ξ_{kj} is the extent of reaction k in stage j and ν_{ki} is the stoichiometric coefficient of component i in reaction k.

The calculation of the extent of the reaction can be considered the reaction rate or chemical equilibrium. If the reaction rate is considered, an expression like the one presented in Eq. (13) is used:

$$\xi_{kj} = V_{hj} k_k \left(\prod_{i=1}^{m} a_{ij}^{(|\nu_{ki}|-\nu_{ki})/2} - \frac{1}{K_k} \prod_{i=1}^{m} a_{ij}^{(|\nu_{ki}|+\nu_{ki})/2} \right) \tag{13}$$

where V_{hj} is the liquid molar holdup in stage j, k_k is the kinetic constant of reaction k, K_k is the chemical equilibrium constant of reaction k and a_{ij} activity of component i on stage j given by Eq. (14):

$$a_{ij} = x_{ij} \gamma_{ij} \tag{14}$$

If the reaction is limited by equilibrium, Eq. (15) is used to obtain the values of the mole fraction to reach the chemical equilibrium:

$$K_k = \prod_{i=1}^{m} \left(x_{ij} \gamma_{ij} \right)^{\nu_{ki}} \exp\left(\frac{P_j - P^{\circ}}{RT_j} \sum_{i=1}^{m} \nu_{ki} V_i^L \right) \tag{15}$$

The chemical equilibrium is reached when the total Gibbs free energy (G) is at its minimum value; in this way, the equilibrium constant is provided by Eq. (16):

$$\ln(K) = -\frac{\Delta G^{\circ}(T)}{RT} \tag{16}$$

One can see that, for equilibrium limited reaction, the extent of reaction is not present in the model equation, but it appears for applying the definition of the extent of reaction in the calculation of the total number of mols of one component based on the total inlet of this component, as shown in Eq. (17):

$$n_{ij} = n_{ij}^{0} + \sum_{k=1}^{p} \nu_{ki} \xi_{kj} \tag{17}$$

where n_{ij} is the quantity of component i in stage j after the reaction reaches the equilibrium based on, n_{ij}^{0}, the total amount of the inlet quantity in the liquid phase (from feed stream or liquid output from stage $j-1$) of component i in stage j. The mole fraction calculated by Eq. (18) must be replaced in Eq. (15)

$$x_{ij} = \frac{n_{ij}}{\sum_{i=1}^{m} n_{ij}} \tag{18}$$

The Murphree tray efficiency, illustrated in **Figure 7** for the McCabe-Thiele method for binary distillation, can be used to better represent the behavior of a mole fraction profile in a reactive distillation column. In this case, the reboiler and condenser are considered in equilibrium, but the tray is considered to have an efficiency compared to the ideal case (vapor-liquid equilibrium). This efficiency can be formulated based on the vapor phase, Eq. (19), or based on the liquid phase, Eq. (19):

$$E_{MV} = \frac{y_{ij} - y_{ij+1}}{y^*_{ij} - y_{ij+1}} \tag{19}$$

$$E_{ML} = \frac{x_{ij} - x_{ij-1}}{x^*_{ij} - x_{ij-1}} \tag{20}$$

where E_{MV} is the Murphree tray efficiency for the vapor phase, E_{ML} is the Murphree tray efficiency for the liquid phase, x_{ij} is the liquid mole fraction of component i that actually leaves stage j, y_{ij} is the vapor mole fraction of component i that actually leaves stage j, x_{ij-1} is the liquid mole fraction of component i that actually leave stage $j-1$ (stage right above j) and goes into stage j, y_{ij+1} is the vapor mole fraction of component i that actually leave stage $j+1$ (stage right bellow j) and goes into stage j, x^*_{ij} is the liquid mole fraction of component i and stage j in equilibrium with the actual vapor composition of stage j $\left(y_{1j}, y_{2j}, \ldots, y_{mj}\right)$ and y^*_{ij} is the vapor mole fraction of component i in stage j in equilibrium with the actual liquid composition of stage j $\left(x_{1j}, x_{2j}, \ldots, x_{mj}\right)$. The relations for y^*_{ij} and x^*_{ij} to the equilibrium are presented, respectively, in Eqs. (21) and (22):

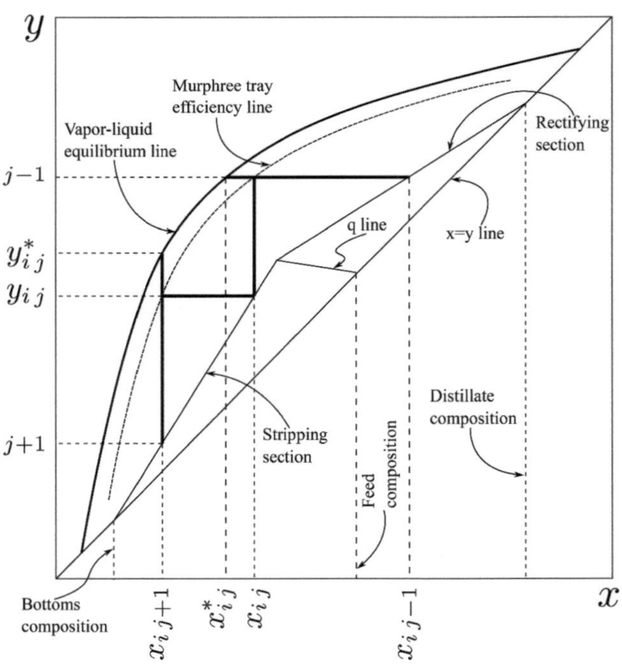

Figure 7.
Schematic representation of Murphree tray efficiency in the MaCabe-Thiele diagram.

$$y_{ij}^{*} = K_{ij}\left(x_{1j}, x_{2j}, \dots, x_{mj}, y_{1j}^{*}, y_{2j}^{*}, \dots, y_{mj}^{*}, T_j, P_j\right) x_{ij} \tag{21}$$

$$y_{ij} = K_{ij}\left(x_{1j}^{*}, x_{2j}^{*}, \dots, x_{mj}^{*}, y_{1j}, y_{2j}, \dots, y_{mj}, T_j, P_j\right) x_{ij}^{*} \tag{22}$$

where P_j is the pressure in stage j.

Another efficiency definition, normally used for avoiding integer optimization (in this case, the stage efficiency can assume continuous variables between 0 and 1), is the bypass efficiency illustrated in **Figure 8**. This concept assumes that only a fraction (proportional to the bypass efficiency of stage j, ε_j) of vapor and liquid inlet streams (L_{j-1} and V_{j+1}) flow into stage j and reach the equilibrium (L_j^* and V_j^*), the other fraction of inlet stream bypass and mix with the equilibrium outlet streams [28].

Sometimes, equilibrium (EQ) stage simulations are denominated to be rigorous, but this term is not totally correct because, in actual operation, the phases in a column stage normally operate a little far from the equilibrium [76]. A more realistic (physically consistent) approach available in the literature for modeling the relation between the compositions of vapor and liquid phases is the non-equilibrium (NEQ) in which the finite mass transfer rates across the vapor-liquid interface are accounted for, that is, a rate-based model. However, the rate-based model is much more complicated than the equilibrium model and also more difficult to converge [67]. For this model, it is assumed that the total amount of mass transfer resistance is in the thin films in the border between the vapor-liquid phases, and the mass transfer in these two films is due to molecular diffusion. In the bulk fluid phases, the mixing level is so high that the composition is almost none [1]. The description of interface mass transfer can be based on Maxwell-Stefan theory for the calculation of heat and mass transfer [1, 8, 41, 58, 67, 77, 78]. Near the vapor-liquid interface, there are two boundary layers (one for each phase) in which the composition is variable, as shown in **Figure 9**, at the vapor-liquid interface can be assumed to be phase equilibrium. In the case this model is considered, energy and molar balances must consider each phase separately [79]. The non-equilibrium mathematical models usually provide more results than the equilibrium models. However, the availability of reliable mass transfer correlations would be a prerequisite for the use of a non-equilibrium stage mode [4].

There are also models for heterogeneous catalytic distillation that take into account simultaneous mass transfer and reaction inside the catalyst particle using the Maxwell-Stefan theory [1, 80].

If it is, necessary or desired, many combinations of rate-based or equilibrium reaction and rate-based mass and heat transfer, phase equilibrium or tray efficiency can be used [69].

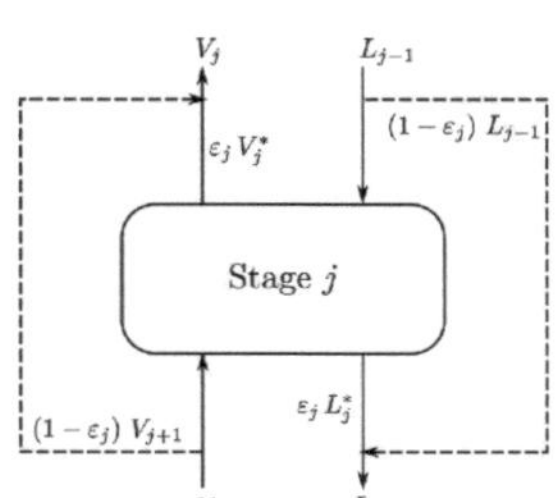

Figure 8.
Illustration of bypass efficiency.

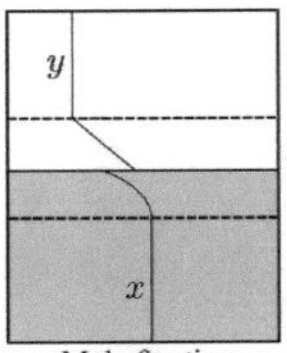

Figure 9.
Schematic representation of a non-equilibrium stage.

The transient model also can be used to study the steady-state solution [4, 24, 69, 76, 81, 82]. In this case, the conservation laws become differential equations, and the solution of this model needs a method for solving the stiffness of differential-algebraic equations (DAE) [1, 32, 76, 82–85]:

$$\frac{d\left(x_{ij}V_{hj}\right)}{dt} = z_{ij}F_j + r_{ij} + y_{ij+1}V_{j+1} + x_{ij-1}L_{j-1} - x_{ij}\left(L_j + U_j\right) - y_{ij}\left(V_j + W_j\right) \quad (23)$$

$$\frac{dV_{hj}}{dt} = F_j + R_j + V_{j+1} + L_{j-1} - \left(L_j + U_j\right) - \left(V_j + W_j\right) \quad (24)$$

$$\frac{d\left(V_{hj}u_j\right)}{dt} = h_j^F F_j + H_{j+1}V_{j+1} + h_{j-1}L_{j-1} - h_j\left(L_j + U_j\right) - H_j\left(V_j + W_j\right) - Q_j \quad (25)$$

where u_j is the internal energy of liquid phase in stage j.

If the liquid molar holdup can be considered as constant, Eqs. (23) and (25) can be simplified as follows:

$$V_{hj}\frac{dx_{ij}}{dt} = z_{ij}F_j + r_{ij} + y_{ij+1}V_{j+1} + x_{ij-1}L_{j-1} - x_{ij}\left(L_j + U_j\right) - y_{ij}\left(V_j + W_j\right) \quad (26)$$

$$V_{hj}\frac{du_j}{dt} = h_j^F F_j + H_{j+1}V_{j+1} + h_{j-1}L_{j-1} - h_j\left(L_j + U_j\right) - H_j\left(V_j + W_j\right) - Q_j \quad (27)$$

There are also in the literature some pseudo-transient mathematical modeling methods that turn the system algebraic eq. (AE) model that describes the process in steady state into a system of differential-algebraic equation, a mathematical model that is statically equivalent to original models [28, 86].

For some cases, molar vapor and liquid flow rates can be considered as constant, in addition to the assumption that there is no heat loss, which eliminates the need for energy balances. Binary distillation is known as the McCabe-Thiele method and can be applied graphically [87]. This principle can be applied to more than two components, and it is known as non-heat effect.

A good mathematical model must represent appropriately the real behavior of the process, and to get this behavior is necessary also a good numerical method for solving the mathematical model. The simulation is the translation of the numerical results obtained by this solution procedure to physical meaning. The next section discusses some methods and strategies used in the simulation of steady-state of reactive distillation.

5. Simulation

The solution process of a mathematical model is known as simulation (nowadays, simulation is synonymous with computational simulation) [68]. Mathematical models used for representing the behavior of some properties relevant to chemical processes are increasingly large due to the inclusion of so many details that aim to reach a realistic description of processes [88].

The complexity of a steady-state reactive distillation model is not due to mass and energy conservation laws, and it comes mainly from constitutive relations of chemical equilibrium or kinetics, mass transfer (in a more simplified way, phase equilibrium), and relations to compute enthalpy [68]. Due to the complexity of the models, it is necessary to apply robust methods or good solution strategies to make it possible to reach solution convergence.

The development of phenomenological mathematical model for a chemical process operating in steady state usually results in a system of non-linear algebraic equations that must be solved iteratively. The model is composed of N algebraic equations, as shown in Eq. (28), and the solution is obtained with respect to a set of N unknown variables, X:

$$F(X) = \begin{bmatrix} f_1(X) \\ f_2(X) \\ \vdots \\ f_n(X) \end{bmatrix} = \begin{bmatrix} f_1(x_1, x_2, x_3, \dots, x_N) \\ f_2(x_1, x_2, x_3, \dots, x_N) \\ \vdots \\ f_n(x_1, x_2, x_3, \dots, x_N) \end{bmatrix} \tag{28}$$

$$X = \begin{bmatrix} x_1 \\ x_2 \\ \vdots \\ x_N \end{bmatrix} \tag{29}$$

The most popular methods for solving this kind of problem are Newton-Raphson and similar ones because these methods present fast convergence when good initial guesses are given, but these methods can fail when no good initial estimates are available, or there are several solutions [75, 89]. The search direction of the Newton-Raphson method is given by the Jacobian matrix, given by Eq. (30):

$$J(X) = \begin{bmatrix} \frac{\partial f_1(X)}{\partial x_1} & \frac{\partial f_1(X)}{\partial x_2} & \dots & \frac{\partial f_1(X)}{\partial x_N} \\ \frac{\partial f_2(X)}{\partial x_1} & \frac{\partial f_2(X)}{\partial x_2} & \dots & \frac{\partial f_2(X)}{\partial x_N} \\ \vdots & \vdots & \ddots & \vdots \\ \frac{\partial f_N(X)}{\partial x_1} & \frac{\partial f_N(X)}{\partial x_2} & \dots & \frac{\partial f_N(X)}{\partial x_N} \end{bmatrix} \tag{30}$$

Using the Newton-Raphson method is a necessary effort to obtain the derivatives analytically, which prevents its use, so it is common to use numerical derivatives. Computer algebra like *Maple* and *Mathematica* can make the analytical derivatives obtaining process easier, but these programs tend to be slow when compared to *Fortran* or *C* [88].

The iterative process of the Newton-Raphson method to approximate the solution of $k+1$ iteration depends on the solution obtained in the last iteration (k), as shown in Eq. (31):

$$X^{(k+1)} = X^{(k)} + \Delta X^{(k)} \tag{31}$$

where $\Delta X^{(k)}$ is obtained by the solution of the linear system of Eq.(32):

$$J\left(X^{(k)}\right)\Delta X^{(k)} = -F\left(X^{(k)}\right) \tag{32}$$

Newton-Raphson method is fast and robust near the solution; however, its performance strongly depends on the choice of good initial estimates [86]. An improvement of the Newton-Raphson method can be obtained by relaxation [1], like the method proposed by [90], Eq. (33), that uses a relaxation parameter s, which is different from the unity when the solution tends to diverge, that is Eq. (34) is not satisfied:

$$X^{(k+1)} = X^{(k)} + s_i^{(k)}\Delta X^{(k)} \tag{33}$$

$$\sqrt{\sum_{j=1}^{N} f_j^2\left(X^{(k)} + s_1^{(k)}\Delta X^{(k)}\right)} < \sqrt{\sum_{j=1}^{N} f_j^2\left(X^{(k)}\right)} \tag{34}$$

where

$$s_2^{(k)} = \frac{\sqrt{1+6\eta}-1}{3\eta} \tag{35}$$

$$\eta = \frac{\sum_{j=1}^{N} f_j^2\left(X^{(k)} + s_1^{(k)}\Delta X^{(k)}\right)}{\sum_{j=1}^{N} f_j^2\left(X^{(k)}\right)} \tag{36}$$

Homotopy method can guarantee that the approximate solution is reached if it is chosen an adequate auxiliary homotopy function, $G(X)$, presented in Eq. (37). The solutions to the auxiliary function may be easily guessed/given/known [89]:

$$G(X) = \begin{bmatrix} g_1(X) \\ g_2(X) \\ \vdots \\ g_N(X) \end{bmatrix} \tag{37}$$

Then, we define the homotopy function as presented in Eq. (38):

$$H(X,\lambda) = (1-\lambda)G(X) + \lambda F(X) \tag{38}$$

where λ is the homotopy parameter that varies from 0 to 1, being that, for $\lambda = 0$ $H(X,0) = G(X)$ and $\lambda = 1$ $H(X,1) = F(X)$.

The goal is to solve $H(X,\lambda) = 0$ instead of $F(X) = 0$, but in this case, there are $N+1$ unknowns and N equation thus for the solution to be possible, the value of λ is fixed, starting in 0, after solving the homotopy function with a fixed value of *lambda*, its value is sequentially changed until the value $\lambda = 1$ is reached. The solution obtained with the last value of λ is used as an initial guess to the next step to avoid the situation of divergence.

The homotopy path from zero to one can be followed by taking sequential steps of small values of λ and by applying a Newton-Raphson method at each step to track the curve; this is called piecewise linear. However, the parametrized λ might lead to poor performance and turning points can be encountered, leading to a failure in the solution procedures. To make the method more efficient, the homotopy function can be turned into an initial value problem (IVP) of a system of ordinary differential equations, and solved by an appropriate numerical method for IVP. The variation in the λ value can have the use of a predictor-corrector continuation method applied [91]. Homotopy-continuation methods are mostly applied in problems that present high difficulty solved by other methods, like Newton-Raphson [92].

Some works have adopted a procedure to apply the Newton-Raphson method to solve, all together, the equations obtained from mass and energy balances, phase equilibrium equations, chemical equilibrium or rates of reaction equations, and any additional equation [88]. Other works have applied the Newton-Raphson method with numerical evaluation of the Jacobian matrix to algorithms that divide the models into subsets of equations [59].

A solution strategy, adapted from conventional distillation, for solving the steady-state reactive distillation modeling is called "tearing equations," shown in **Figure 10a**, in which, after some rearrangements into the model equations, the solution is divided

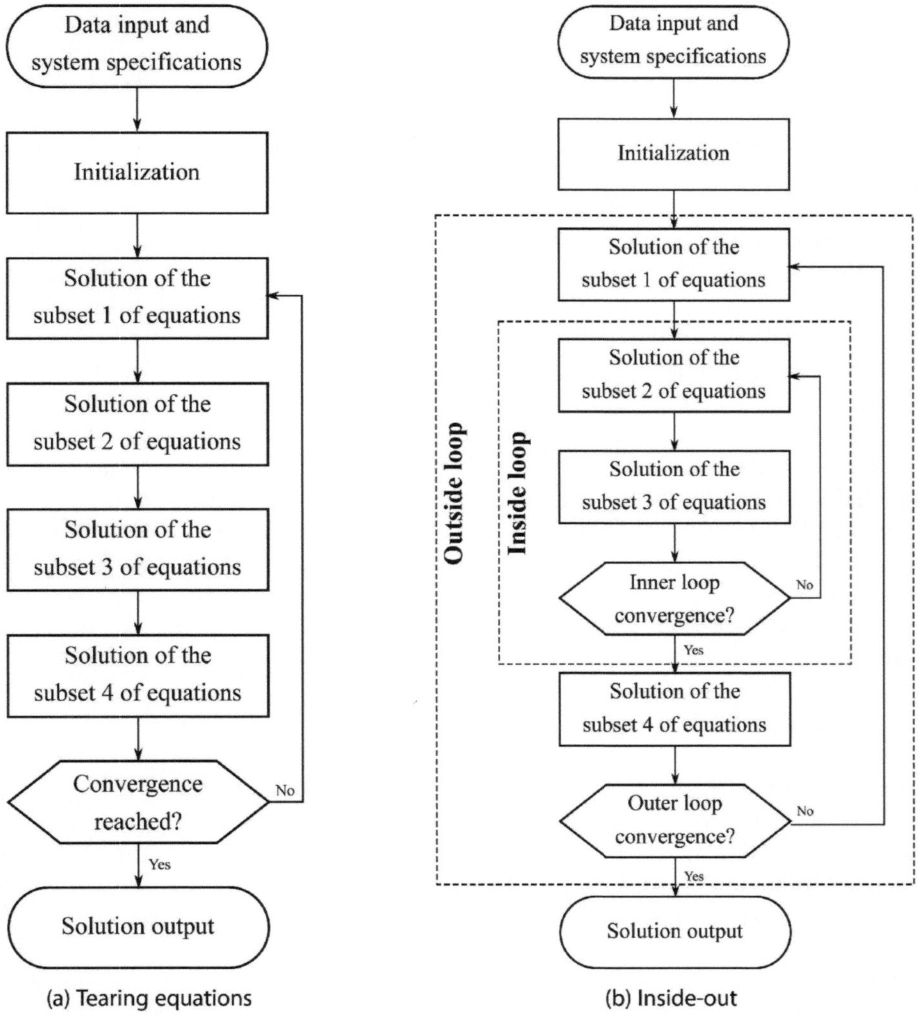

Figure 10.
Flowchart of the solution algorithms.

into many steps, a solution of a subset of unknowns in each step, in a way that it is possible to avoid the solution of non-linear equations sets for all unknowns, mostly the subsets of equations are composed only of linear equations because the other unknowns (not being calculated in the current step) are considered constant [75, 93, 94]. This strategy can make the implementation of the solution algorithm simpler and the solution faster. The subset of equations sequence is arranged in a generic way because the sequence can vary from one method to another and mainly because the strategy of subset grouping can be different, but the subset of unknowns must be contained in the subset of equation, for example, it is possible to use the component mass balance to calculate the mass fraction, use the mass balance in each stage to calculate the liquid stream leaving each stage, the energy balance to calculate the vapor stream leaving each stream, use the chemical equilibrium or reaction rate to calculate the extent of reaction, the summation of molar fractions combined with phase equilibrium to calculate the temperatures.

An improvement in the "tearing equations" methodology is to add an inner iterative process, **Figure 10b**, in which just some of the subsets are solved. This can reduce the number of iterations in the outer loop and make the solution even faster, called "inside-out" [74, 95, 96]. The subsets of equations for these methods normally are tridiagonal systems of linear equations, which can be solved by, for example, Thomas algorithm to reduce the computer time needed.

Similar to tearing methods, the simultaneous correction method also separates the equations in many subsets, but the equations are first linearized [96]. Most algorithms for solving reactive distillation steady state models are adaptations of well-known algorithms for solving conventional distillation steady state models, like the Naphtali-Sandholm method; in this method, equations are grouped by stages rather than by components and solved by the Newton-Raphson method [46, 97, 98]. Also, dynamic fuzzy neural networks have been implemented to solve mathematical models of steady-state engineering process that are represented by a system of non-linear equations [99].

The model equations can be implemented using a programming language (like Fortran or C), model builders (like gPROMS, an equation-oriented simulator) or a sequential modular simulator. The use of sequential modular simulators like Aspen Plus, Aspen HYSYS, CHEMCAD, ProSimPlus, etc., is common for analyses of reactive distillation systems [2, 4, 10, 13, 21–23, 26, 27, 36, 37, 43, 44, 46, 61, 65, 67, 100–115]. Due to a wide variety of libraries for equipments that they include, thus allowing the users to simulate almost any process plant [116]. Most research about the conceptual design of the reactive distillation process is still carried out on commercial simulators such as Aspen Plus for steady-state design [14].

Aspen (Advanced System for Process Engineering) Plus has been used for many recent process modeling involving physical chemistry, chemical thermodynamics, mass and energy balances, and chemical reaction engineering [117–119]. This software offers a completely integrated solution to chemical process industries, making it possible to use it in many tasks of process engineering, from design to capital analysis. It has built-in model libraries from many chemical process, including distillation and reactive distillation [44]. It has a strong thermodynamics database and robust numerical solvers [120].

There are in the steady state simulation Aspen Plus robust modules for equilibrium stage and rate-based models. RADFRAC module is based upon a rigorous equilibrium stage model for solving MESH equations, while RATEFRAC is a rate-based model [32, 46, 67, 108, 109, 114, 119, 121–124].

HYSYS Software is a powerful engineering simulation tool comprised of various components that provide an extremely powerful approach to steady-state modeling [125].

Although the steady-state modeling and simulation of reactive distillation column has been studied for decades, there are still some challenges and opportunities in this research subject, among which some are presented in the next section.

6. Challenges and opportunities

One can see that the reactive distillation process has received many proposals for modification in the hardware structure, and many others can be proposed. Also, there are many study cases reported in the literature, and many others will be studied. However, the greatest challenge that we can see is the solution procedure of the steady-state model.

The availability of many commercial simulators in which the user does not need to know what equations are solved in the simulation is a very interesting tool because the researcher can focus only on the process being studied, but it is important to know what is being solved in a computational simulation to correctly understand the results being reached. Also, the modular sequential simulators can be used to simulate almost any process plant, but there is a disadvantage that arises when a specific process, whose behavior cannot be accurately represented by any of these general models, is present in a process plant. In this case, if the researcher is not capable of obtaining and solving a mathematical model by using programming languages, he/she cannot do any simulation about the process, and there are some characteristics and strategies in the solution procedures that can be improved.

The solution of a steady-state model of a reactive distillation problem involves the solution of a system of highly non-linear algebraic equations [75, 93]. The solution of a system of non-linear equations is very hard and usually needs good initial guesses in the way the method presents convergence [126]. For a steady-state reactive distillation model, a good initialization strategy is required due to the highly non-linear behavior of the thermodynamic model and the kinetic equations [14]. There are very few works that have some propose to obtain the initial estimates of a steady state of a reactive distillation process [75, 93]. Some authors use the solution obtained in other works as initial estimates [88, 127] or use the solution of a more simple model and the results [85].

A methodology for obtaining an initial guess is provided for the solution of steady-state modeling of equilibrium distillation by [94] and steady-state modeling of equilibrium reactive distillation by [75, 93]. These authors used a non-heat effect for flow rates, a linear profile for temperature and a quadratic profile for the extent of the reaction [75, 93]. The quadratic profile for the extent of reaction with the maximum in the feed stage was also used by [74]. The variables already obtained as initial guess are enough to start the numerical solution procedure that uses a tearing equations strategy. For the initial guess of the temperature profile [74, 95], a linear interpolation for the temperature considering the bubble point in the condenser and the dew point in the reboiler dew point for an average composition of all feed streams. The authors in [95] do not present details on the equations, but they also cited that vapor and liquid-phase fractions are initialized by assuming constant molar overflow in the column and also considered ideal equilibrium ratios and non-reactive systems for other initial guesses.

The existence of multiple steady state (MSS) makes the importance of initial estimates even greater. Simulation studies suggest that reactive distillation processes exhibit complex multiple steady-state behavior in many cases because of many physicochemical phenomena interacting deliberately [33, 34, 47, 72, 82, 85, 128–131]. There is the possibility of using commercial packages to identify the MSS, like ASPEN PLUS [37, 100, 132, 133], HYSYS [18, 73, 78, 134, 135], and SPeedUp™ [136]. Reactive distillation can exhibit input multiplicity and output multiplicity. The input multiplicity occurs when the same output is obtained by a multiple set of inputs, and an output multiplicity occurs when the same input results in multiple sets of outputs. Among those can exist stable and unstable states. To design a reactive distillation process, it is important to discover all steady states. The multiple solutions can be reached by performing a sensitivity analysis in some parameters or varying the initial estimates [37, 137, 138].

Baharev and Neumaier [139] proposed a method to find all steady-state solutions of the distillation column. In this work, it is stated that there is no need for initial estimates. The method requires a specific but fairly general block-sparsity pattern, with a linear growth in the computation effort. The algorithm requires that its input system of equations has been permuted in such a way that the Jacobian is in lower block Hessenberg form. The variables are partitioned into subvectors, and the equations are divided into equation subsets in a way that only variables from the subvectors can appear in the subsets of equations, but one variable can appear in more than one subvector. The original problem is reduced to much smaller subproblems. There is some lack of information in that proposal because it is stated that there is no need for initial estimates, but some subset of equations must be non-linear, which makes it necessary to have initial estimates.

Homotopy method can be applied without the supply of good initial estimates, but it turn the computational effort much greater, because the system of equations must be modified and solved many times.

After all, one can see that the best method, considering implementation difficulty, computational effort, and success in obtaining the solution, is the use of methods like Newton-Raphson or Broyden with good initial estimates exploring the sparsity of the model like used in simultaneous correction, Naphtali-Sandholm method, tearing equations strategy or inside-out algorithm, or even solving all equation together.

Heuristic methods, for global optimization, have been receiving some interest in the last years [140]. Advances in computing and information technology allow chemical engineers to solve complex design problems [25] in such a way the genetic algorithm has been used to solve simple problems that are a system of algebraic equations [141], for one just can write the objective function as the sum of the absolute value (or the square) of each algebraic function ($F(X) = 0$):

$$F_{\text{obj}} = \sum_i |f_i(X)| \tag{39}$$

$$F_{\text{obj}} = \sum_i (f_i(X))^2 \tag{40}$$

The use of heuristic optimization methods may require much computational time to solve large systems of algebraic equation, but it can quickly find some good candidates as initial estimates for a fast deterministic numerical solution method like

Newton-Raphson or Broyden. And, the nature of heuristic methods can be used to generate different initial estimates to be used in the study of multiple steady states. Also, the method proposed by [75, 93] can be used as one of the possible solution at the start of the heuristic algorithm.

Another important opportunity is about the representation of the generic flows in a column stage because it is a smart choice to make the stage configuration representation as generic as possible in a way that just one model can be used to represent all the stages in a column, being necessary only the specification of the streams connections, this also makes it easier to write the computational code. Many authors used their models generic stages in a way that can be applied to the stage presented in **Figure 6**. That is a good choice, but its representation and resulting material and energy balances cannot be used, for example, in all elements of columns that have side reactor, side columns, pump around, etc. So, here we are proposing a more generic stage schematic model, presented in **Figure 11**, that can be used to represent all these. The change is that the side liquid and flow rates can be sent out of the process (product output stream) for side stages (side reactor, heat exchanger and side column stage) or even for another stage in the same column. It turns the model more complex and with many more specifications needed, but makes possible to generate all material and energy balances in the same way, even for the side reactor, pump around, or side columns, which require a considerable amount of attention on the stages numbering and specifications. Even the bypass efficiency can be modeled with this flow scheme, but the column stages must be numbered sequentially from top to bottom, and the other elements (side column, side reactors, pump around, etc.) must be numbered out of this range.

For this schematic representation, the global mass balance is represented by Eq. (41):

$$F_j + R_j + V_{j+1} + \sum_{\substack{l=1 \\ l \neq j}}^{n} \omega_{lj} W_l + L_{j-1} + \sum_{\substack{l=1 \\ l \neq j}}^{n} \lambda_{lj} U_l - (L_j + U_j) - (V_j + W_j) = 0 \quad (41)$$

$$F_j^L + F_j^V + R_j + + \sum_{\substack{l=1 \\ l \neq j}}^{n} \tau_{lj} V_l + \sum_{\substack{l=1 \\ l \neq j}}^{n} \sigma_{lj} V_l - L_j - V_j - P_j^L - P_j^V = 0 \quad (42)$$

where ω_{lj} is the fraction of side vapor flow rate that leaves stage l and goes into stage j, λ_{lj} is the fraction of side liquid flow rate that leaves stage l and goes into stage j,

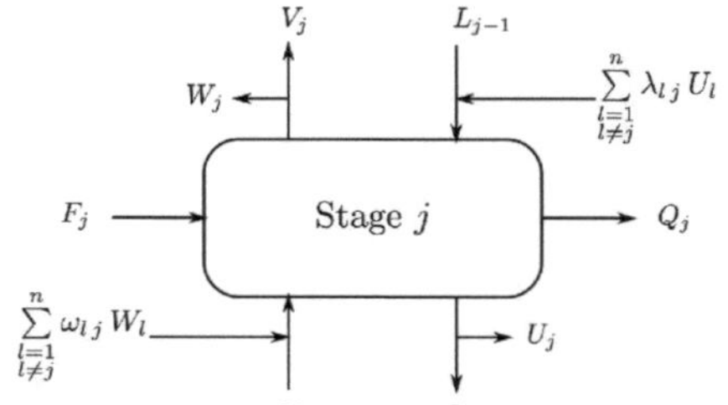

Figure 11.
Schematic of a more generic stage in a distillation column.

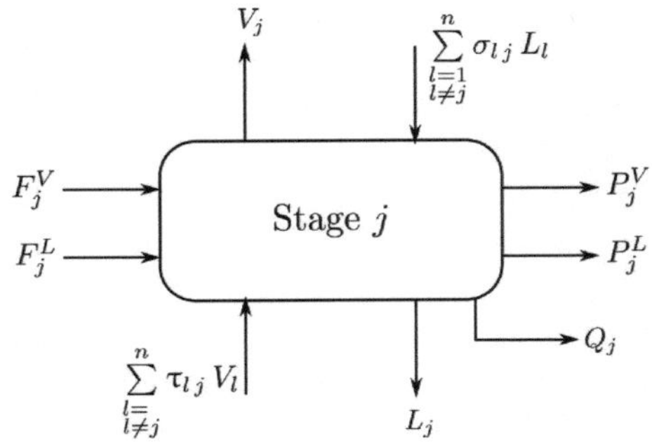

Figure 12.
Schematic of another generic stage in a distillation column.

ω_{jj} is the fraction of side vapor flow rate that leaves the process (product output stream) from stage j and λ_{jj} is the fraction of side liquid flow rate that leaves the process (product output stream) from stage j.

In another generic stage modeling we are proposing, as shown in **Figure 12**, it is considered separated liquid and vapor feed streams (F_j^L and F_j^V), as same as product streams (P_j^L and P_j^V) in a way that it is not considered side stream, like in the previous model, but the liquid and vapor outputting stage j can go into any other stage. Thus, the stages can be numbered freely, even column side elements. For this schematic representation, the global mass balance is represented by Eq. (42).where τ_{lj} is the fraction of vapor flow rate that leaves stage l and goes into stage j, σ_{lj} is the fraction of liquid flow rate that leaves stage l and goes into stage j, F_j^L is the feed flow rate in the liquid phase in stage j, F_j^V is the feed flow rate in the vapor phase in stage j, P_j^L is the product flow rate in the liquid phase from stage j and P_j^V is the product flow rate in the vapor phase from stage j.

7. Conclusions

This work was carried out as a systematic literature review about modeling and simulation of steady-state reactive distillation processes. Using the *Methodi Ordinatio*, a portfolio was obtained composed of 144 papers because it eliminated the articles that resulted in a negative *Ordinatio* index. With the bibliometric article's data, it was possible to find the countries, journals, and authors with the most publications in the portfolio, and it was also possible to see the temporal evolution of the number of articles. Reactive distillation is an evolution of process in which the reaction and separation take place in different units, but there are many modifications in the configuration of the reactive distillation columns aiming to improve the product quality or reduce the capital or operational costs, being that the reactive dividing wall column can be highlighted and there are many modifications in a specific case that are being evaluated. A steady-state simulation is a fundamental tool for evaluating a change in the process without effectively doing the modification in practice, lowering the cost in analysis, and doing the modification only if it is favorable. Lately, these simulations are mainly conducted by commercial simulators, like Aspen Plus, which are excellent tools that greatly reduce the investigation time for the process modification. There are some strategies that can be improved in the modeling and in the solution of the models, mainly about the initial estimates, a very important component in the solution process that has few proposes in the literature; it also can be possible to propose some more general flow schemes in and out of a stage in a way that only one general balance can be applied in the main column stages and side elements like side reactor, pump around and side columns.

Conflict of interest

The authors declare no conflict of interest.

Abbreviations

AE	algebraic equation
CD	catalytic distillation
DAE	differential-algebraic equation
DWC	dividing wall column
EQ	equilibrium
eRDWC	enzymatic reactive dividing wall column
HIDiC	heat-integrated distillation column
HIRDiC	heat-integrated reactive distillation column
HP	heat pump
HPRD	heat pump reactive distillation
IVP	initial value problem
MSS	multiple steady states
NEQ	Non-equilibrium
PSD	pressure swing distillation
PSRD	pressure swing reactive distillation
RD	reactive distillation
RDC	reactive dividing column
RDWC	reactive dividing wall column
REDWC	reactive-extractive dividing wall column
SRC	side reaction column
SS	steady state
TCDS	thermally coupled distillation sequences
VC	vapor compression column
RC	vapor recompression column
Symbols	
a_{ij}	activity of component i on stage j
E_{ML}	Murphree tray efficiency for liquid phase
E_{MV}	Murphree tray efficiency for vapor phase
F	system of non-linear algebraic equation
F_j	feed flow rate to stage j
F_j^L	feed flow rate to stage j in the liquid phase
F_j^V	feed flow rate to stage j in the vapor phase
G	Gibbs-free energy
G°	Gibbs-free energy in standard state
h_j	enthalpy of liquid stream flow from stage j

h_j^F	enthalpy of feed stream flow to stage j
H_j	enthalpy of vapor stream flow from stage j
J	Jacobian matrix
K_{ij}	liquid-vapor equilibrium relation for the component i on stage j
k_k	kinetic constant of reaction k
K_k	chemical equilibrium constant of reaction k
L_j	liquid flow rate from stage j
L_j^*	liquid flow rate from stage j in equilibrium
n	total number of stages
n_{ij}	quantity of component i in stage j after the reaction reached the equilibrium (for the extent of reaction calculation)
n_{ij}^0	total amount of the inlet quantity in the liquid phase (from feed stream or liquid output from stage $j-1$) of component i in stage j (for the extent of reaction calculation)
m	total number of components
Q_j	heat load from stage j
P	pressure
P_j	pressure in stage j
P_j^L	product flow rate from stage j in the liquid phase
P_j^V	product flow rate from stage j in the vapor phase
P°	pressure in standard state
P^{sat}	saturation pressure
r_{ij}	generation rate of component i on stage j
R	ideal gas constant
R_j	total rate of generation on stage j
T	temperature
T_j	temperature of stage j
u_j	internal energy of liquid phase in stage j
U_j	liquid side flow rate from stage j
V_j	vapor flow rate from stage j
V_j^*	vapor flow rate from stage j in equilibrium
V_{hj}	liquid molar holdup in stage j
V_i^L	component i volume in liquid phase
W_j	vapor side flow rate from stage j
x_{ij}	liquid mole fraction of component i in stage j
x_{ij}^*	equilibrium liquid mole fraction of component i in stage j
X	set of unknowns
y_{ij}	vapor mole fraction of component i in stage j
y_{ij}^*	equilibrium vapor mole fraction of component i in stage j
z_{ij}	feed mole fraction of component i in stage j
α_i	relative volatility of component i
γ_{ij}	activity coefficient of component i on stage j
δ_j	binary variable used to identify reactive stages, equal to 1 if the reaction occurs in stage j, 0 otherwise
ε_j	bypass efficiency of stage j
η	reduction in the Euclidean norm of the set of non-linear functions between two steps of a Newton-Raphson relaxation method

η_j^L	ratio between the liquid side stream and the liquid stream outputting stage J and inputting the near stage $j+1$
η_j^V	ratio between the vapor side stream and the vapor stream outputting stage J and inputting the near stage $j-1$
λ_{lj}	fraction of side liquid flow rate that leaves stage l and goes into stage $j\nu_{ki}$stoichiometric coefficient of component i in reaction k
ξ_{kj}	extent of reaction k in stage j
σ_{lj}	fraction of vapor flow rate that leaves stage l and goes into stage j
τ_{lj}	fraction of liquid flow rate that leaves stage l and goes into stage j
$\hat{\phi}_{ij}$	fugacity coefficient of component i on stage j
ϕ_i^{sat}	fugacity coefficient of pure component i in saturation state
ω_{lj}	fraction of side vapor flow rate that leaves stage l and goes into stage j

Author details

Vilmar Steffen[1]*, Maiquiel Schmidt de Oliveira[2] and Edson Antonio da Silva[3]

1 Academic Departments of Engineering (DAENG), Federal University of Technology – Parana (UTFPR), Rua Gelindo João Folador, Francisco Beltrão, Paraná, Brazil

2 Academic Department of Physics, Statistics and Mathematics (DAFEM), Federal University of Technology – Parana (UTFPR), Rua Gelindo João Folador, Francisco Beltrão, Paraná, Brazil

3 Center for Engineering and Exact Sciences, State University of Paraná, Rua da Faculdade, Toledo, Paraná, Brazil

*Address all correspondence to: vilmars@utfpr.edu.br

References

[1] Kenig EY, Górak A, Pyhälahti A, Jakobsson K, Aittamaa J, Sundmacher K. Advanced rate-based simulation tool for reactive distillation. AICHE Journal. 2004;**50**(2):322-342

[2] Li J, Zhou H, Sun L, Zhang N. Design and control of different pressure thermally coupled reactive distillation for synthesis of isoamyl acetate. Chemical Engineering and Processing – Process Intensification. 2019;**139**:51-67

[3] Plesu AE, Bonet J, Plesu V, Bozga G, Galan MI. Residue curves map analysis for tert-amyl methyl ether synthesis by reactive distillation in kinetically controlled conditions with energy-saving evaluation. Energy. 2008;**33**(10): 1572-1589

[4] Fernandez MF, Barroso B, Meyer XM, Meyer M, Le Lann MV, Le Roux GC, et al. Experiments and dynamic modeling of a reactive distillation column for the production of ethyl acetate by considering the heterogeneous catalyst pilot complexities. Chemical Engineering Research and Design. 2013; **91**(12):2309-2322

[5] Grüner S, Mohl KD, Kienle A, Gilles ED, Fernholz G, Friedrich M. Nonlinear control of a reactive distillation column. Control Engineering Practice. 2003;**11** (8):915-925

[6] Masuku CM, Xiaojun L, Hildebrandt D, Glasser D. Reactive distillation in conventional Fischer-Tropsch reactors. Fuel Processing Technology. 2015;**130** (C):54-61

[7] Mario Mihal', Zuzana Svandova, and Jozef Markos. Steady state and dynamic simulation of a hybrid reactive separation process. Chemical Papers. 2010;**64**(2):193-202

[8] Noeres C, Kenig EY, Górak A. Modelling of reactive separation processes: Reactive absorption and reactive distillation. Chemical Engineering and Processing. 2003;**42**(3): 157-178

[9] Olanrewaju MJ, Al-Arfaj MA. Impact of disturbance magnitudes and directions on the dynamic behavior of a generic reactive distillation. Chemical Engineering and Processing: Process Intensification. 2006;**45**(2):140-149

[10] Athimathi S, Radhakrishna TK. Control system design for a single feed ETBE reactive distillation column. Chemical Engineering and Technology. 2006;**29**(10):1137-1154

[11] Bock H, Wozny G, Gutsche B. Design and control of a reaction distillation column including the recovery system. Chemical Engineering and Processing: Process Intensification. 1997;**36**(2):101-109

[12] Bogatykh I, Hoffmann C, Kozachynskyi V, Illner M, Osterland T, Wilharm T, et al. Insights into dynamic process intensification for reactive distillation columns. Chemical Engineering and Processing – Process Intensification. 2022;**177**:108978

[13] Chen C-L, Chung Y-H, Lee H-Y. Design and control of reactive distillation process for the production of methyl valerate. Industrial and Engineering Chemistry Research. 2016; **55**(5):1347-1360

[14] Chung Y-H, Peng T-H, Lee H-Y, Chen C-L, I-Lung Chien. Design and control of reactive distillation system for esterification of levulinic acid and n-butanol. Industrial and Engineering

Chemistry Research. 2015;**54**(13): 3341-3354

[15] Giwa A, Adeyi AA, Adeyi VA. Cascade PID control of a reactive distillation process for biodiesel production: A comparison with conventional PID control. International Journal of Engineering Research in Africa. 2018;**35**:134-144

[16] Sakhre V. Reactive distillation: Modeling, simulation, and optimization. In: Distillation – Modelling, Simulation and Optimization. London, UK: IntechOpen; 2019

[17] Kharaji S, Sadeghi J, Shahraki F, Khalilipour MM. A new control structure for tert-amyl methyl ether production using reactive distillation. ISA Transactions. 2020;**97**:53-66

[18] Khaledi R, Young BR. Modeling and model predictive control of composition and conversion in an ETBE reactive distillation column. Industrial and Engineering Chemistry Research. 2005; **44**(9):3134-3145

[19] Cossio-Vargas E, Barroso-Muñoz FO, Hernandez S, Segovia-Hernandez JG, Irene Cano-Rodriguez M. Thermally coupled distillation sequences: Steady state simulation of the esterification of fatty organic acids. Chemical Engineering and Processing: Process Intensification. 2012;**62**:176-182

[20] Delgado-Delgado R, Hernández S, Barroso-Muñoz FO, Segovia-Hernández JG, Castro-Montoya AJ. From simulation studies to experimental tests in a reactive dividing wall distillation column. Chemical Engineering Research and Design. 2012;**90**(7):855-862

[21] Hernández S, Sandoval-Vergara R, Barroso-Muñoz FO, Murrieta-Dueñas R, Hernández-Escoto H, Segovia-Hernández JG, et al. Reactive dividing wall distillation columns: Simulation and implementation in a pilot plant. Chemical Engineering and Processing: Process Intensification. 2009;**48**(1): 250-258

[22] Gor NK, Mali NA, Joshi SS. Intensified reactive distillation configurations for production of dimethyl ether. Chemical Engineering and Processing – Process Intensification. 2020;**149**:107824

[23] Holtbruegge J, Wierschem M, Lutze P. Synthesis of dimethyl carbonate and propylene glycol in a membrane-assisted reactive distillation process: Pilot-scale experiments, modeling and process analysis. Chemical Engineering and Processing: Process Intensification. 2014; **84**:54-70

[24] Al-Arfaj M, Luyben WL. Comparison of alternative control structures for an ideal two-product reactive distillation column. Industrial and Engineering Chemistry Research. 2000;**39**(9):3298-3307

[25] Almeida-Rivera CP, Grievink J. Process design approach for reactive distillation based on economics, exergy, and responsiveness optimization. Industrial and Engineering Chemistry Research. 2008;**47**(1):51-65

[26] Carlos Cárdenas-Guerra J, Susana Figueroa-Gerstenmaier J, Reyes-Aguilera A, Hernández S. Simulation study of a reactive distillation process for the ethanol production. Chemical Engineering Transactions. 2018;**69**: 613-618

[27] Dhale AD, Myrant LK, Chopade SP, Jackson JE, Miller DJ. Propylene glycol and ethylene glycol recovery from aqueous solution via reactive distillation.

Chemical Engineering Science. 2004;**59**(14):2881-2890

[28] Ma Y, Luo Y, Yuan X. Equation-oriented optimization of reactive distillation systems using pseudo-transient models. Chemical Engineering Science. 2019;**195**:381-398

[29] Machado G, Castier M, dos Santos M, Nishiyama F, Aranda D, Cardozo-Filho L, et al. Reactive distillation applied to biodiesel production by esterification: Simulation studies. In: Distillation Processes – From Solar and Membrane Distillation to Reactive Distillation Modelling, Simulation and Optimization. London, UK: IntechOpen; 2022

[30] Mohl KD, Kienle A, Gilles ED, Rapmund P, Sundmacher K, Hoffmann U. Nonlinear dynamics of reactive distillation processes for the production of fuel ethers. Computers and Chemical Engineering. 1997;**21**(SUPPL.1): S989-S994

[31] Patan AK, Mekala M, Thamida SK. Simulation of a steady-state continuous catalytic reactive distillation column by using a multiscale capillary model. Chemical Engineering and Technology. 2022;**45**(5):878-889

[32] Pilavachi PA, Schenk M, Perez-Cisneros E, Gani R. Modeling and simulation of reactive distillation operations. Industrial & Engineering Chemistry Research. 1997;**36**(8): 3188-3197

[33] Reder C, Gehrke V, Marquardt W. Steady state multiplicity in esterification distillation columns. Computers & Chemical Engineering. 1999;**23**(SUPPL. 1):S407-S410

[34] Scenna NJ, Benz SJ. Start-up operation of reactive columns with multiple steady states: The ethylene glycol case. Industrial and Engineering Chemistry Research. 2003;**42**(4): 873-882

[35] Lalu Seban V, Kirubakaran BK, Roy, and T K Radhakrishnan. GOBF-ARMA based model predictive control for an ideal reactive distillation column. Ecotoxicology and Environmental Safety. 2015;**121**:110-115

[36] Mingyuan H, Tian H. Design of process and control scheme for cyclohexanol production from cyclohexene using reactive distillation. Chinese Journal of Chemical Engineering. 2021;**40**:96-105

[37] Yang B, Jiang W, Zhao G, Wang H, Shiqing L. Multiplicity analysis in reactive distillation column using ASPEN PLUS. Chinese Journal of Chemical Engineering. 2006;**14**(3): 301-308

[38] Novita FJ, Zou BY, Lee HY. Design and control of a hybrid reactive extraction configuration for the production of tert-butyl alcohol. Journal of Cleaner Production. 2019;**239**:118018

[39] Agirre I, Barrio VL, Güemez B, Cambra JF, Arias PL. Catalytic reactive distillation process development for 1,1 diethoxy butane production from renewable sources. Bioresource Technology. 2011;**102**(2):1289-1297

[40] Al-Arfaj MA, Luyben WL. Design and control of an olefin metathesis reactive distillation column. Chemical Engineering Science. 2002;**57**(5):715-733

[41] Baur R, Higler AP, Taylor R, Krishna R. Comparison of equilibrium stage and nonequilibrium stage models for reactive distillation. Chemical Engineering Journal. 2000;**76**(1):33-47

[42] Bezzo F, Bertucco A, Forlin A, Barolo M. Steady-state analysis of an industrial reactive distillation column. Separation and Purification Technology. 1999;**16**(3):251-260

[43] Iftakher A, Mansouri SS, Nahid A, Tula AK, Shoukat Choudhury MA, Lee JH, et al. Integrated design and control of reactive distillation processes using the driving force approach. AICHE Journal. 2021;**67**(6):e17227

[44] Anene RC, Giwa A. Modelling, simulation and sensitivity analysis of a fatty acid methyl ester reactive distillation process using aspen plus. International Journal of Engineering Research in Africa. 2016;**27**:36-50

[45] Cheng JK, Lee HY, Huang HP, Cheng Ching Y. Optimal steady-state design of reactive distillation processes using simulated annealing. Journal of the Taiwan Institute of Chemical Engineers. 2009;**40**(2):188-196

[46] Ram Singh MV, Kumar P, Kaistha N. Steady-state reactive distillation simulation using the naphtali-sandholm method. Canadian Journal of Chemical Engineering. 2007;**85**(1):75-82

[47] Sneesby MG, Tadé MO, Smith TN. Mechanistic interpretation of multiplicity in hybrid reactive distillation: Physically realizable cases. Industrial and Engineering Chemistry Research. 1998;**37**(11):4424-4433

[48] Steffen V, Schmidt M, de Oliveira C, Brusamarello Z, Trojan F. A new normalized index for ranking papers in systematic literature reviews. Decision Analytics Journal. 2024;**10**:100439

[49] De Oliveira MS, Steffen V, Trojan FV. Systematic literature review on electric vehicles and multicriteria decision making: Trends, rankings, and future perspectives. Journal of Intelligent Management Decision. 2024;**3**:22-41

[50] Liberati A, Altman DG, Tetzlaff J, Mulrow C, Gøtzsche PC, Ioannidis JPA, et al. The prisma statement for reporting systematic reviews and meta-analyses of studies that evaluate healthcare interventions: Explanation and elaboration. BMJ. 2009;**339**:W-65-W-94

[51] Ensslin L, Dutra A, Ensslin SR, Chaves LC. Vinicius Dezem, and others. Research process for selecting a theoretical framework and bibliometric analysis of a theme: Illustration for the management of customer service in a bank. Modern Economy. 2015;**6**(06):782

[52] Kosztyán ZT, Csizmadia T, Katona AI. SIMILAR – Systematic iterative multilayer literature review method. Journal of Informetrics. 2021;**15**(1): 101111

[53] Pagani RN, Kovaleski JL, Resende LM. Methodi ordinatio: A proposed methodology to select and rank relevant scientific papers encompassing the impact factor, number of citation, and year of publication. Scientometrics. 2015;**105**(3):2109-2135

[54] Pagani RN, Kovaleski JL, Martins LM, de Resende. TICs na composição da Methodi Ordinatio: construção de portfólio bibliográfico sobre Modelos de Transferência de Tecnologia. Ciência da Informação. 2018;**24**(2):161-187

[55] Schmidt M, de Oliveira V, Steffen, and Flavio Trojan. A systematic review of the literature on video assistant referees in soccer: Challenges and opportunities in sports analytics. Decision Analytics Journal. 2023;**7**: 100232

[56] Schmidt M, de Oliveira V, Steffen AC, de Francisco, and Flavio Trojan.

Integrated data envelopment analysis, multi-criteria decision making, and cluster analysis methods: Trends and perspectives. Decision Analytics Journal. 2023;**8**:100271

[57] Alvarez-Ramirez J. Singular reactive flash dynamics. Chemical Engineering and Processing: Process Intensification. 2013;**69**:119-125

[58] Higler AP, Taylor R, Krishna R. The influence of mass transfer and mixing on the performance of a tray column for reactive distillation. Chemical Engineering Science. 1999;**54**(13–14): 2873-2881

[59] Bo C, Tang J, Yangjin Bai X, Qiao LD, Zhang S. The design and control of distillation column with side reactors for chlorobenzene production. Chinese Journal of Chemical Engineering. 2012; **20**(6):1113-1120

[60] Pan Q, Li J, Shang X, Ma S, Liu J, Sun M, et al. Controllability, energy-efficiency, and safety comparisons of different control schemes for producing n-butyl acetate in a reactive Dividing Wall column. Industrial and Engineering Chemistry Research. 2019;**58**(22): 9675-9689

[61] Gao X, Yang Y, Chen M, Cheng Q, Kairui L. Novel heat pump reactive distillation and dividing-wall column reactive distillation processes for synthesizing isopropyl acetate to save TAC and reduce CO2 emissions. Chemical Engineering and Processing – Process Intensification. 2022;**171**:108746

[62] Liang J, Zhou H, Li J, Kong W, Ma Z, Sun L. Comparison of dynamic performances for heat integrated reactive distillation considering safety. Chemical Engineering and Processing – Process Intensification. 2021;**160**:108294

[63] Jana AK. Heat integrated distillation operation. Applied Energy. 2010;**87**(5): 1477-1494

[64] Vanaki A, Eslamloueyan R. Steady-state simulation of a reactive internally heat integrated distillation column (R-HIDiC) for synthesis of tertiary-amyl methyl ether (TAME). Chemical Engineering and Processing: Process Intensification. 2012;**52**:21-27

[65] Gao X, Chen M, Yang Y, Cheng Q, Kairui L, Li Y. Improved design and optimization of synthetic n-butyl acetate process to save energy. Separation Science and Technology (Philadelphia). 2022;**57**(6):1000-1011

[66] Liu J, Wan G, Dong M, Kong J, Yang W, Han S, et al. Dynamic controllability strategy of reactive-extractive dividing wall column for the separation of water-containing ternary azeotropic mixture. Separation and Purification Technology. 2023;**304**:122338

[67] Peng J, Lextrait S, Edgar TF, Bruce R, Eldridge. A comparison of steady-state equilibrium and rate-based models for packed reactive distillation columns. Industrial and Engineering Chemistry Research. 2002;**41**(11):2735-2744

[68] Steffen V. Introductory chapter: Distillation. In: Distillation – Modelling, Simulation and Optimization. London, UK: IntechOpen; 2019

[69] Chin SY, Mohamed AR, Ahmad AL, Bhatia S. Esterification of palmitic acid with iso-propanol in a catalytic distillation column: Modeling and simulation studies. International Journal of Chemical Reactor Engineering. 2006; **4**(1). DOI: 10.2202/1542-6580.1235

[70] Labovský J, Švandová Z, Markoš J, L'udovít Jelemenský. Model-based HAZOP study of a real MTBE plant.

Journal of Loss Prevention in the Process Industries. 2007;**20**(3):230-237

[71] Sneesby MG, Tadé MO, Datta R, Smith TN. ETBE synthesis via reactive distillation. 1. Steady-state simulation and design aspects. Industrial and Engineering Chemistry Research. 1997; **36**(5):1855-1869

[72] Sneesby MG, Tadé MO, Smith TN. Steady-state transitions in the reactive distillation of MTBE. Computers and Chemical Engineering. 1998;**22**(7–8): 879-892

[73] Zuzana Švandová, Juraj Labovský, Jozef Markoš, and L'udovít Jelemenský. Impact of mathematical model selection on prediction of steady state and dynamic behaviour of a reactive distillation column. Computers and Chemical Engineering. 2009;**33**(3): 788-793

[74] Wang L, Sun X, Xia L, Wang J, Xiang S. Inside-out method for simulating a reactive distillation process. PRO. 2020;**8**(5):604

[75] Steffen V, Silva EA. Numerical methods and initial estimates for the simulation of steady-state reactive distillation columns with an algorithm based on tearing equations methodology. Thermal Science and Engineering Progress. 2018;**6**:1-13

[76] Katariya AM, Kamath RS, Moudgalya KM, Mahajani SM. Non-equilibrium stage modeling and non-linear dynamic effects in the synthesis of TAME by reactive distillation. Computers and Chemical Engineering. 2008;**32**(10):2243-2255

[77] Kotora M, Švandová Z, Markoš J. A three-phase nonequilibrium model for catalytic distillation. Chemical Papers. 2009;**63**(2):197-204

[78] Z. Švandová, J. Markoš, and L' Jelemenský. Impact of mass transfer coefficient correlations on prediction of reactive distillation column behaviour. Chemical Engineering Journal. 2008;**140** (1–3):381–390

[79] Baur R, Taylor R, Krishna R. Influence of column hardware on the performance of reactive distillation columns. Catalysis Today. 2001;**66**(2–4): 225-232

[80] Arnoud Higler R, Krishna, and Ross Taylor. Nonequilibrium modeling of reactive distillation: A dusty fluid model for heterogeneously catalyzed processes. Industrial and Engineering Chemistry Research. 2000;**39**(6):1596-1607

[81] Huss RS, Chen F, Malone MF, Doherty MF. Reactive distillation for methyl acetate production. Computers and Chemical Engineering. 2003;**27**(12): 1855-1866

[82] Katariya AM, Moudgalya KM, Mahajani SM. Nonlinear dynamic effects in reactive distillation for synthesis of TAME. Industrial and Engineering Chemistry Research. 2006;**45**(12): 4233-4242

[83] Reepmeyer F, Repke J-U, Wozny G. Analysis of the start-up process for reactive distillation. Chemical Engineering and Technology. 2003;**26** (1):81-86

[84] Schenk M, Gani R, Bogle IDL, Pistikopoulos EN. A hybrid approach for reactive separation systems. Computers & Chemical Engineering. 1999;**23** (SUPPL. 1):S419-S422

[85] Schenk M, Gani R, Bogle D, Pistikopoulos EN. A hybrid modelling approach for separation systems involving distillation. Chemical

Engineering Research and Design. 1999; **77**(6):519-534

[86] Li K, Luo Y, Yuan X. Equation-oriented optimization of reaction distillation column considering tray hydraulics. Separation and Purification Technology. 2022;**295**:121229

[87] Mccabe WL, Thiele EW. Graphical Design of Fractionating Columns. Industrial & Engineering Chemistry. 1925;**17**(6):605-611

[88] Alfradique MF, Castier M. Modeling and simulation of reactive distillation columns using computer algebra. Computers and Chemical Engineering. 2005;**29**(9):1875-1884

[89] Tzong-Mou W. A study of convergence on the newton-homotopy continuation method. Applied Mathematics and Computation. 2005; **168**(2):1169-1174

[90] Broyden CG. A class of methods for solving nonlinear simultaneous equations. Mathematics of Computation. 1965;**19**(92):577-593

[91] Tanskanen J, Pohjola VJ. A robust method for predicting state profiles in a reactive distillation. Computers & Chemical Engineering. 2000;**24**(1):81-88

[92] Taylor R, Krishna R. Modelling reactive distillation. Chemical Engineering Science. 2000;**55**(22): 5183-5229

[93] Steffen V, da Silva EA. Steady-state modeling of reactive distillation columns. Acta Scientiarum. Technology. 2012;**34**(1):61-69

[94] Steffen V, da Silva EA. Steady-state modeling of equilibrium distillation. In: Distillation – Innovative Applications and Modeling. London, UK: InTech; 2017

[95] Khaledi R, Bishnoi PR. A method for Modeling two- and three-phase reactive distillation columns. Industrial & Engineering Chemistry Research. 2006; **45**(17):6007-6020

[96] Simandl J, Svrcek WY. Extension of the simultaneous-solution and inside-outside algorithms to distillation with chemical reactions. Computers and Chemical Engineering. 1991;**15**(5): 337-348

[97] Christiansen LJ, Michelsen ML, Fredenslund A. Naphtali-sandholm distillation calculations for ngl mixtures near the critical region. Computers & Chemical Engineering. 1979;**3**(1):535-542

[98] Naphtali LM, Sandholm DP. Multicomponent separation calculations by linearization. AICHE Journal. 1971;**17** (1):148-153

[99] Francisco J. Sanchez-Ruiz. Reactive distillation modeling using artificial neural networks. In: Steffen V, editor. Distillation Processes, Chapter 8. London, UK, Rijeka: IntechOpen; 2022

[100] Ramzan N, Faheem M, Gani R, Witt W. Multiple steady states detection in a packed-bed reactive distillation column using bifurcation analysis. Computers and Chemical Engineering. 2010;**34**(4):460-466

[101] Alharthi MA. Exergy analysis of reactive distillation coupled with high-pressure column for the synthesis of dimethyl carbonate. PRO. 2022;**10**(6): 1219

[102] Brehelin M, Forner F, Rouzineau D, Repke JU, Meyer X, Meyer M, et al. Production of n-propyl acetate by reactive distillation. Experimental and

theoretical study. Chemical Engineering Research and Design. 2007;**85**(1 A): 109-117

[103] Carlos Cárdenas-Guerra J, López-Arenas T, Hernández S, Hernández-Escoto H, Figueroa-Gerstenmaier S, Reyes-Aguilera JA. Non-linear dynamics simulation study of a reactive distillation process for the ultra-low sulfur diesel production. Chemical Engineering Transactions. 2017;**57**:1681-1686

[104] Giwa A, Giwa SO. Steady-state modelling and simulation of a reactive distillation process for n-butyl acetate production using CHEMCAD. International Journal of Engineering Research in Africa. 2017;**29**:70-80

[105] González DR, Bastidas P, Rodríguez G, Gil I. Design alternatives and control performance in the pilot scale production of isoamyl acetate via reactive distillation. Chemical Engineering Research and Design. 2017; **123**:347-359

[106] Díaz VHG, Willis MJ. Ethyl acetate production from dilute bioethanol with low energy intensity. Journal of Cleaner Production. 2022;**376**:134137

[107] Holtbruegge J, Heile S, Lutze P, Górak A. Synthesis of dimethyl carbonate and propylene glycol in a pilot-scale reactive distillation column: Experimental investigation, modeling and process analysis. Chemical Engineering Journal. 2013;**234**:448-463

[108] Huang Z, Li J, Wang L, Jiang H, Qiu T. Novel procedure for the synthesis of dimethyl carbonate by reactive distillation. Industrial and Engineering Chemistry Research. 2014;**53**(8): 3321-3328

[109] Kaymak DB. Design and control of an alternative process for biobutanol purification from ABE fermentation. Industrial and Engineering Chemistry Research. 2019;**58**(5):1957-1965

[110] Mo Y, Dong L, Zhang C, Quan H. Process study and simulation for the recovery of 1,1,2,2,3,3,4-heptafluorocyclopentane by reactive distillation. PRO. 2022;**10**(6):1146

[111] Orjuela A, Kolah A, Lira CT, Miller DJ. Mixed succinic acid/acetic acid esterification with ethanol by reactive distillation. Industrial and Engineering Chemistry Research. 2011;**50**(15): 9209-9220

[112] Orjuela A, Kolah A, Hong X, Lira CT, Miller DJ. Diethyl succinate synthesis by reactive distillation. Separation and Purification Technology. 2012;**88**:151-162

[113] Santoso H, Bao J, Lee PL. Operability analysis of MTBE reactive distillation column using a process simulator. Chemical Product and Process Modeling. 2009;**4**(3). DOI: 10.2202/1934-2659.1376

[114] Weinfeld JA, Bruce Eldridge R, Owens S. Evaluation of the aldol condensation of propionaldehyde as a reactive Dividing Wall column test system. Industrial and Engineering Chemistry Research. 2022;**61**(23): 8220-8232

[115] Bor Yih Y, Tseng TY, Yang ZY, Shen SJ. Evaluation on the solketal production processes: Rigorous design, optimization, environmental analysis, and control. Process Safety and Environmental Protection. 2022;**157**: 140-155

[116] Domancich AO, Perez V, Hoch PM, Brignole NB. Systematic generation of a CAPE-OPEN compliant simulation module from GAMS and FORTRAN

models. Chemical Engineering Research and Design. 2010;**88**(4):421-429

[117] An W, Lin Z, Chen J, Zhu J. Simulation and analysis of a reactive distillation column for removal of water from ethanol-water mixtures. Industrial and Engineering Chemistry Research. 2014;**53**(14):6056-6064

[118] Dohare RK, Ansari P. Heat integration of reactive divided wall distillation column. In: Distillation Processes – From Solar and Membrane Distillation to Reactive Distillation Modelling, Simulation and Optimization. London, UK: IntechOpen; 2022

[119] Singh D, Gupta RK, Kumar V. Simulation of a plant scale reactive distillation column for esterification of acetic acid. Computers and Chemical Engineering. 2015;**73**:70-81

[120] Ghosh S, Srinivas S. Reactive distillation for methanol synthesis: Simulation-based design methodology. Korean Journal of Chemical Engineering. 2022;**39**(9):2291-2306

[121] Bhatia S, Ahmad AL, Mohamed AR, Chin SY. Production of isopropyl palmitate in a catalytic distillation column: Experimental studies. Chemical Engineering Science. 2006;**61**(22): 7436-7447

[122] Chen Y, Ding R, Wang Y, Ye Y, Zhang G. Kinetics study and process simulation of reactive distillation for the synthesis of ?-caprolactone. Canadian Journal of Chemical Engineering. 2021; **99**(4):999-1010

[123] Guo B, Li Y. Analysis and simulation of reactive distillation for gasoline alkylation desulfurization. Chemical Engineering Science. 2012;**72**: 115-125

[124] Kiss AA. Heat-integrated reactive distillation process for synthesis of fatty esters. Fuel Processing Technology. 2011;**92**(7):1288-1296

[125] Karacan S, Karacan F. Steady-state optimization for biodiesel production in a reactive distillation column. Clean Technologies and Environmental Policy. 2015;**17**(5):1207-1215

[126] De Castilhos F, Corazza JV, de Oliveira, and Marcos Lúcio Corazza. Application of a subdivision algorithm for solving nonlinear algebraic systems. Acta Scientiarum. Technology. 2008;**30**:5

[127] Medina IJM, Montiel-Maldonado C. Technical re-conversion of an MTBE plant to produce ETBE. International Journal of Chemical Reactor Engineering. 2007;**5**(1):1-13

[128] Gehrke V, Marquardt W. A singularty theory approach to the study of reactive distillation. Computers & Chemical Engineering. 1997;**21**(S): S1001-S1006

[129] Mohl K-D, Kienle A, Gilles E-D. Multiple steady states in a reactive distillation column for the production of the fuel ether TAME I: Theoretical analysis. Chemical Engineering and Technology. 1998;**21**(2):133-136

[130] Rapmund P, Sundmacher K, Hoffmann U. Multiple steady states in a reactive distillation column for the production of the fuel ether TAME part II: Experimental validation. Chemical Engineering and Technology. 1998;**21** (2):136-139

[131] Yamaki T, Matsuda K. Control of reactive distillation through the multiple steady state conditions. Chemical Engineering Transactions. 2012;**29**: 223-228

[132] Yamaki T, Matsuda K, Na-Ranong D, Matsumoto H. Energy-saving performance of reactive distillation process for TAME synthesis through multiple steady state conditions. Chemical Engineering and Processing – Process Intensification. 2018;**130**: 101-109

[133] Yamaki T, Matsuda K, Na-Ranong D, Matsumoto H. Intensification of reactive distillation for TAME synthesis based on the analysis of multiple steady-state conditions. PRO. 2018;**6**(12):241

[134] Švandová Z, Markoš J, Jelemenský L'U. Influence of the reactive distillation column configuration on its performance: A computational study. Chemical Papers. 2008;**62**(1):58-64

[135] Švandová Z, Markoš J. Theoretical study on transesterification in a combined process consisting of a reactive distillation column and a pervaporation unit. Chemical Papers. 2011;**65**(2):167-176

[136] Sneesby MG, Tade MO, Smith TN. Reaction hysteresis: A new cause of output multiplicity in reactive distillation. Developments in Chemical Engineering and Mineral Processing. 1999;**7**(1–2):41-56

[137] Rodríguez IE, Zheng A, Malone MF. The stability of a reactive flash. Chemical Engineering Science. 2001;**56**(16): 4737-4745

[138] Sharifzadeh M. Implementation of a steady-state inversely controlled process model for integrated design and control of an ETBE reactive distillation. Chemical Engineering Science. 2013;**92**:21-39

[139] Baharev A, Neumaier A. A globally convergent method for finding all steady-state solutions of distillation columns. AICHE Journal. 2014;**60**(2): 410-414

[140] Steffen V. Particle swarm optimization with a simplex strategy to avoid getting stuck on local optimum. AI, Computer Science and Robotics Technology. 2022;**1–40**:2022

[141] Panwar H, Singh D, Singh A. Genetic algorithm for solving simple mathematical equality problem. International Research Journal of Engineering and Technology. 2020;7: 7622-7627

Chapter 6

Evaluation of the Best Operating Conditions in Distillation Columns: A Case Study for the Separation between Nonylphenol and Dinonylphenol

Julio Cesar Ribeiro Nunes, Ricardo de Freitas Fernandes Pontes, Fabio Rodolfo Miguel Batista and Rafael M. Matricarde Falleiro

Abstract

Nonylphenol is a very important product for the chemical industry due to its widespread use as a base for several other chemical products. Among the major industrial sectors that use nonylphenol is the production of non-ionic surfactants, which are used from the home and personal care industry to the agrochemical industry. This study aims to define the optimal or quasi-optimal operating conditions for the separation system, which is composed of packed columns. Using simulation tools, the best operating conditions are attained, and the dinonylphenol subproduct generation is minimized. The commercial simulator Aspen Plus® was used for this study as the analysis tool for the mentioned objectives. The developed model is validated with technical data, measures are taken in a nonylphenol plant, and parameters are used in the same plant. From the obtained data from the plant, the best process performance is evaluated regarding cost-benefit analysis and safety concerns. The study shows a potential to reduce the subproduct production by 30% and the reboilers' heat loads by 2%.

Keywords: process simulation, Aspen Plus® software, nonylphenol, distillation column, packed column

1. Introduction

Nonylphenol is a very important reactant in the surfactant industry, and its main application is in ethoxylated non-ionic surfactants, which use about 80% of the produced nonylphenol [1]. Nonylphenol is produced by a reaction between phenol, nonene, and a basic catalyst. The various nonylphenol applications range from the home and personal care industry to the agrochemical industry.

In chemical terms, nonylphenol is an alkylphenol, hence a phenol derivative where one more hydrogen atom in the aromatic ring is replaced by alkyl radicals. The most important alkylphenols are composed of alkyl radicals that have 2–12 carbon atoms. Besides their use as non-ionic surfactants, alkylphenols are also used as phenolic resins, polymeric additives, and agrochemicals. Nonylphenol is produced from an alkene (nonene) by acid catalysis (Lewis acid or ion-exchange resin), ensuring the replacement of hydrogen by the alkyl radical in the aromatic ring [1, 2]. The overall reaction is highly exothermic (−23.7 kcal/gmol) and reversible as presented in **Figure 1**.

The overall reaction process synthesizes the isomers ortho-nonylphenol and para-nonylphenol, and the latter is the major product. Besides the main product, the process also produces dinonylphenol as a subproduct, and this occurs by the reaction between one nonylphenol molecule with one nonene molecule. Dinonylphenol is much less used in the chemical industry compared to nonylphenol. Hence, an optimal operating point must be defined to reduce the dinonylphenol production to the minimum possible value because if this subproduct is produced in relatively large quantities, then the plant profit is reduced.

A productive process understanding is required to achieve nonylphenol synthesis optimization. Such understanding is done by theoretical studies that allow technical analysis and plant operation planning. Simulation software is used to evaluate the process under several different conditions without interfering in the plant operation. Moreover, by using simulation software, the optimization studies can be better directed saving time and resources that would be spent if the studies were directed toward a non-optimal configuration.

1.1 Nonylphenol production process

Figure 2 presents the nonylphenol productive process used by a major chemical plant in southeastern Brazil. This particular process consists of the following three major unit operations: reaction, phenol recovery, and nonylphenol purification.

The phenol alkylation reaction occurs in two serial mixture atmospheric reactors and is filled with an acid ionic exchange resin bed. Reactor 1 (R-1) operates within an 80–130°C range, and Reactor 2 (R-2) operates within an 85–135°C range. The temperature increase is gradual and based on the accumulated production of 1°C for each 300 ton of produced nonylphenol [3]. The reaction feed stream contains a phenol excess. The major raw material cost comes from the nonene purchase; hence nonene is the limiting reactant. The feed stream contains trace amounts of water from the phenol and nonane. The overall formation selectivity is 96.3% nonylphenol, 2.7% dinonylphenol, and 1% heavier compounds [2].

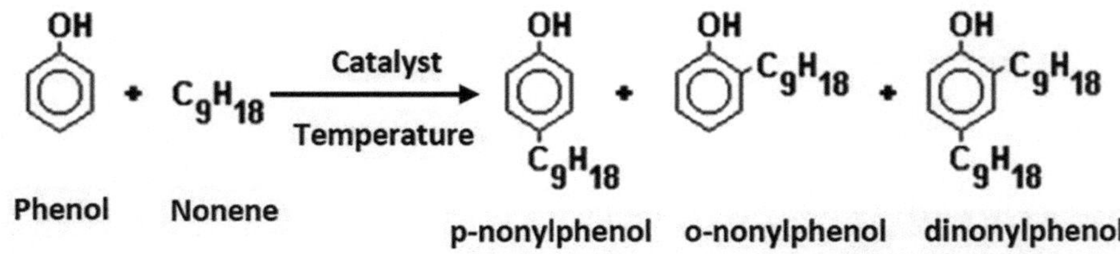

Figure 1.
Nonylphenol synthesis overall reaction.

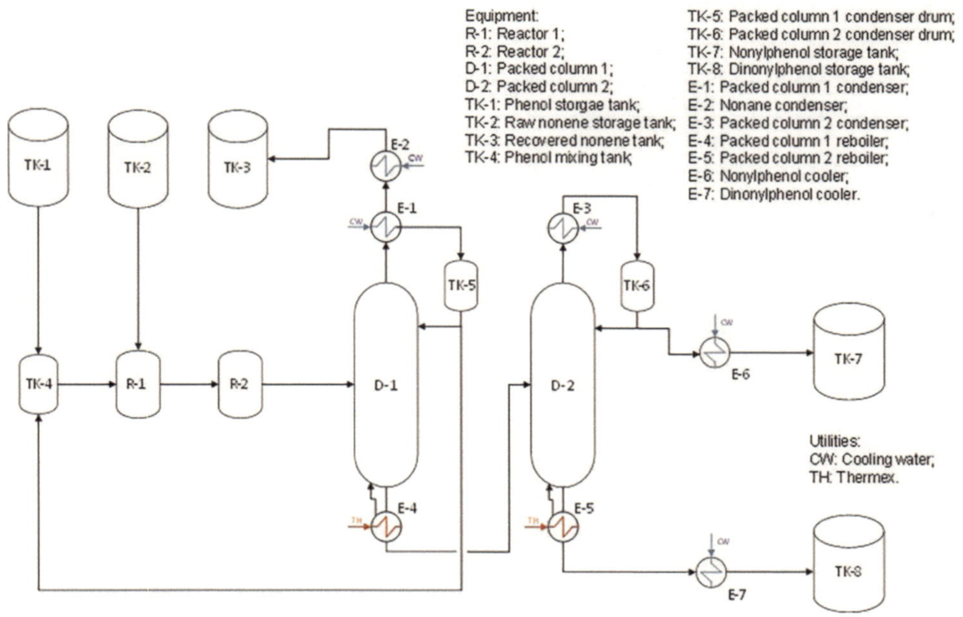

Figure 2.
Process flowchart for the nonylphenol production process.

The process is catalyzed by an acid ionic exchange resin that replaces a Lewis acid. The use of the resin catalyst reduces the raw material demand, yields less subproduct production, and allows better control of mixing parameters. On the other hand, using resin catalyst requires increased capital costs, and the catalyst activity decreases with time. Therefore, reaction control must account for the activity reduction to ensure efficient reactors' operation.

1.2 Phenol recovery

As the name suggests, this stage aims for the recovery of the unreacted phenol. Distillation is the used separation method. A column (D-1) with structured packing (SULZER BX 50) separates phenol from the heavier compounds in the Reactor 2 (R-2) effluent. The column operating pressure ranges from 0.240 to 0.267 bar. The estimated maximum pressure drop from the bottom to the top is about 0.027 bar. Column D-1 temperature profile lies within the 135°C (top) to 256°C (bottom) range.

Column D-1 feed stream is composed of nonylphenol, dinonylphenol, phenol, nonene, nonane, and traces of water. Since nonane lowers the nonylphenol yield in the reaction process, Column D-1 operates with a partial condenser to separate nonane from the unreacted phenol. The vapor distillate stream is mostly composed of nonane, nonene, and water, while the liquid distillate stream is phenol-rich. The vapor distillate stream is condensed in the Nonane Condenser (E-2), and the outgoing condensed stream goes to storage. The liquid distillate stream is recycled to Tank TK-4 at the beginning of the process, so the recovered phenol is reused in the reaction.

Column D-1 bottom stream is composed of heavier compounds, mainly nonylphenol, and dinonylphenol. As the bottom stream does not meet commercial purity requirements for nonylphenol product, the bottom stream goes to Column D-2 to separate nonylphenol from dinonylphenol.

1.3 Nonylphenol purification

The final stage aims, as previously mentioned, the attainment of nonylphenol products complying with market specifications (minimum 95% purity). The purification is done in Column D-2, which uses structured packing (SULZER BX 50). This column operates at a high vacuum (0.067–0.073 bar). Therefore, the pressure drop in the column must be limited to a maximum of 0.007 bar. Column D-2 operating temperature range is from 216°C (top) to 277°C (bottom).

Nonylphenol at market specification is produced at Column D-2 distillate stream. The purified nonylphenol is then cooled in Exchanger E-6 and stored in Tank TK-7. Therefore, the bottom stream is mainly composed of dinonylphenol, which is also cooled (Exchanger E-7) and stored (Tank TK-8).

2. Packed column and process simulation

In the 1950s, chemical industries used more frequently packed columns as distillation equipment [4]. Usually, packed columns are used for the separation of mixtures with high relative volatility difference, hence mixtures that are relatively easy to separate [5].

Packed column interiors are basically completely filled with a packing, which is either packed or random. Unlike a plate column, there is no space between stages, and the only empty spaces in the column are at the top, the bottom, and around the feed stream inlet(s).

Random packing is made up of small components such as Rasching and pall rings. These components have defined dimensions and geometric shapes. The column interior is filled with packing but in a random manner. Hence, the packing distribution varies randomly with the column height.

While random packing does not have a definite design, structured packing is designed to be the same throughout the stage height. Structured packing is usually formed by perforated metallic sheets that have corrugated sections, in order to create multiple parallel channels in the packing. The structured packing design forces the liquid and vapor streams to flow in countercurrent through the channels forming a liquid film in the packing surface. In this manner, contact between the liquid and vapor is maximized in the structured packing [6].

The main downside of structured packing is the need for a high feed stream flow rate, otherwise, the liquid film formation throughout the column transversal area is compromised. Consequently, low feed stream flow rates can reduce the interphase contact in the column resulting in low-efficiency stages [7].

Compared to structured packing, random packing has a lower capital cost and is installed more easily. However, structured packing is more efficient regarding distillation and imposes a lower head loss. Moreover, depending on the random packing component distribution, a preferential path can be formed, short-circuiting the column, and drastically reducing column efficiency [8, 9].

In packed columns, other internals are essential to avoid a non-uniform distribution, a preferential path formation, or even liquid flowing through the column's internal wall. Among these internals, the main ones are the distributors and re-distributors [8].

Considering the nonylphenol production process, packed columns are fundamental for safe and large-scale production as shown in **Figure 1**. However, as many thorough *locus* studies, process optimization finds problems due to plant operation requirements. To perform any study test, plant production can be seriously reduced or even halted, yielding a profit loss. Additionally, relatively drastic alterations in process conditions can compromise the ability of the plant to return to its original operating conditions. Consequently, process simulators can provide tools to investigate different operating conditions, and these simulators are nowadays widely used in chemical industries.

Many process simulators are only able to evaluate a process under a steady state. However, some process simulators can predict the system behavior in a transient state, and this means that the simulator can predict the system behavior when a perturbation occurs. For plants that have different operating regimes, meaning that the process must migrate from one steady state condition to a different one, some transient state process simulators can perform real-time optimization. Such resources enable simulated studies a greater versatility, and a more rigorous evaluation and closer to the system's actual response. This is crucial for the financial assessment of the productive process [10].

However, the results attained by a process simulator are only as good as the process model built by the engineers responsible for the process simulation. Therefore, the simulation engineers must have a thorough technical knowledge of the process. This means selecting an accurate thermodynamic model must be selected, the adequate unit operations involved in the process, and also arranging these unit operations to create a coherent productive process. Simulation engineers must also correctly interpret the simulation results, in order to ensure that the simulated model predicts the actual operation with precision.

Nowadays, several chemical process simulators are commercially available or even available for free. Aspen Plus® software from Aspen Tech is one of the main available process simulators. This software has an extensive data bank that includes component parameters and unit operations. Hence, this data bank allows Aspen Plus® to achieve accurate simulation results for many processes, specifically for distillation.

3. Case study: nonylphenol separation and purification

The case study focuses on the nonylphenol-production separation system. Therefore, the study encompasses both the phenol recovery-packed column (D-1) and the nonylphenol purification-packed column (D-2). The system was simulated using Aspen Plus® and is shown in **Figure 2**.

For the study development, technical data was obtained from an actual nonylphenol production unit. The data contains information related to the equipment, process operating parameters, and stream composition data, including feed streams' composition and flow rate. With this information, the flowsheet is built in the Aspen Plus® software. For all the different evaluated conditions, optimization is performed to minimize the dinonylphenol production.

3.1 Simulation development

The first step in building the simulation model involved obtaining actual stream data, specifically Column D-1 feed stream and Column D-2 distillate stream. Additionally, the operating parameters for both columns were investigated. A 3-year time interval for data gathering was established from March 2016 to March 2019. The

Component	D-1 feed stream	D-2 distillate stream
Nonylphenol, %p	50–70	95–100
Dinonylphenol, %p	0–5	0–3
Phenol, %p	20–40	0–1
Nonene, %p	0–15	0–1.5
Water, %p	0–0.5	0–0.2

Table 1.
Specification limits for separation system sampling points.

historical data was supplied by the nonylphenol plant. From the data, the average and the standard deviation were calculated, in addition to assessing superior and inferior limits. Only steady-state operation data was considered, so plant startup and shutdown operation data were disregarded. **Table 1** presents the maximum and minimum values for Column D-1 feed stream and Column D-2 distillate stream.

The ideal stream composition is defined to attain the case study objective. This is done by evaluating the periods when dinonylphenol was minimum. The plant production data is available in monthly intervals; thus 3 months were selected as the minimum dinonylphenol-production periods.

3.2 Thermodynamic model

The definition of the selected thermodynamic model for the nonylphenol production system simulation is a crucial step for attaining accurate simulation results. The component's thermodynamic, kinetic, and transport properties are evaluated using the adequate thermodynamic model [11]. In this manner, the component mixture thermodynamic relations and liquid-vapor equilibrium condition in the columns are precisely calculated, hence yielding accurate outgoing stream calculations.

Considering that the compounds present in the process are majorly very polar, that no electrolytes are present, and that the operating pressure is lower than 10 bar, the NRTL model was selected. Besides being the appropriate model in the decision tree given by the simulator, previous studies performed by the plant also used the NRTL model, and the results from these previous studies were deemed accurate.

3.3 Distillation column operating data validation

With the selection of the NRTL thermodynamic model, simulations were performed with Aspen Plus® using the columns' operating date from the unit process flow diagrams (PFD) and equipment data sheet. As these documents also contain the unit material balance for the design condition, this means that the simulation results can be compared with the material balance results.

The packed columns are controlled by the bottom temperature as specified by design. Therefore, this variable was defined as a parameter that must be met by the simulation results. Another important validation point is compliance with the expected outgoing stream composition. In this manner, the nonylphenol in Column D-2 distillate stream must also be met by the simulation results.

For the simulation validation, two situations were considered: comparative evaluation with design conditions and with actual operating conditions obtained from plant instrumentation. For the design scenario, Column D-2 feed and distillate stream compositions are obtained by using the unit material balance, the packed columns' top and bottom temperatures, the reboilers' heat loads, and the hot utility flow rate used in these reboilers. For the actual operating scenario, the distillate stream composition is compared with the stream analytical composition measurements, and also by the actual column top and bottom temperature, as well as the measured column temperature profile.

3.4 Operating conditions evaluation

Two operating strategies are feasible to achieve dinonylphenol reduction, which is the study optimization objective. The first strategy consists in reducing the dinonylphenol in the reaction stage. The second strategy is for Column D-2 to produce a distillate stream (nonylphenol product) with the maximum allowable dinonylphenol content. This strategy can be interpreted as contaminating the nonylphenol product to the limit established by market specifications.

Using the 3-month low dinonylphenol yield, the average operating condition values were used as simulation input parameter values. In this manner, it is possible to compare the simulation results of the product outgoing streams for different flow rates. The results define the feed flow rate range that complies with the required production targets, and also the results show how dinonylphenol production varies according to the feed flow rate.

The simulation results also show how the reboilers' heat loads vary with the feed flow rate. Hence, the hot utility (Thermex) consumption and the resulting operating costs are also evaluated.

3.5 Stream average properties and compositions

Tables 2 and **3** show the main values for the stream operating variables (input and output) for the nonylphenol purification stage. The input variables are used for the

Description	Average Jan/2017	Average Jun/2018	Average Mar/2019
Product outgoing flow rate (nonylphenol) (kg/h)	1519	1279	1077
D-2 top pressure (bar abs)	0.0733	0.0733	0.0725
D-2 head loss from bottom to top (bar)	0.0046	0.0026	—
D-2 top temperature (°C)	211	211	210
D-2 bottom temperature (°C)	261	262	263
D-1 feed flow rate (kg/h)	3535	3030	2713
D-1 top pressure (bar abs)	0.333	0.333	0.333
D-1 head loss from bottom to top (bar)	0.018	0.015	0.012
D-1 feed temperature (°C)	158	152	155
D-1 top temperature (°C)	135	132	136
D-1 bottom temperature (°C)	254	256	257

Table 2.
Average operating parameter values.

Stream	Component	Average Jan/17	Average Jun/18	Average Mar/19
D-1 feed	Nonylphenol, %p	51.13	49.32	48.32
	Dinonylphenol, %p	1.50	1.64	1.76
	Phenol, %p	35.14	30.28	30.71
	Nonene, %p	12.22	18.76	19.21
D-2 distillate	Nonylphenol, %p	98.01	97.55	97.20
	Dinonylphenol, %p	1.38	1.99	2.23
	Phenol, %p	0.27	0.40	0.52
	Nonene, %p	0.34	0.06	0.06

Table 3.
Average composition for D-1 feed stream and D-2 distillate stream (disregarding water content).

simulation model configuration, and the output variables are compared with the simulation results, so the results can be validated.

The average composition shown in **Table 3** for Column D-1 feed stream was used as simulation input variables. Column D-2 distillate stream composition values are compared with the simulation results. The latter stream average flow rate is calculated for the result comparison. The table shows the production results for the period between 2017 and 2019. Notably, the nonylphenol product stream flow rate shows a decline tendency, since nonylphenol demand has also decreased. For the 3-year period, the average nonylphenol product flow rate is 1195 kg/h (**Table 4**).

3.6 Separation system validation

Table 5 data were obtained from the plant project data. Reflux ratio and reboiler heat input data were obtained from the unit design specifications.

Although the actual Column D-1 condenser is a partial one, the simulation considered a total condenser for simplification purposes. The reason for the use of a partial condenser in column D-1 is for the nonane removal, which was not an objective of the study.

The used structured packing is Sulzer BX 50. The packing was inserted above and below the feed stage. The technical data used in the simulation was also obtained from the unit design data. For Column D-1, the upper part (rectification) packing has a 0.387 m diameter and encompasses Stages 2–4. This part of the column is 1.535 m in height and filled with packing. As for the lower part (stripping), the packing has 0.66 m diameter, and measures 1.345 m of height contemplating Stages 5–7. For Column D-2, the whole vessel has a 0.641 m de diameter. The rectification section is

Year	Nonylphenol production (t)	Average hourly production (kg/h)
2017	9650	1469
2018	7728	1176
2019*	6183	941

**2019 considers the production from January to March, and the projection for the remainder of the year.*

Table 4.
Nonylphenol average production.

Input	D-1	D-2
Number of stages	8	6
Condenser type	Total	Total
Reboiler type	Kettle	Kettle
Feed stage	5	4
Column pressure	0.240 bar abs	0.0667 bar abs
Hot utility	Thermex	Thermex
Structured packing	Sulzer BX 50	Sulzer BX 50
Reflux ratio	0.1	0.1
Reboiler heat load	343 Mcal/h	146 Mcal/h

Table 5.
Simulation input data for packed columns D-1 and D-2.

Parameter	Value
Latent heat of vaporization (kcal/kg)	68.0
Incoming temperature (°C)	280
Outgoing temperature (°C)	280
Overall heat transfer coefficient (kcal/h.m^2.°C)	1282

Table 6.
Thermex (hot utility) properties.

composed of Stages 2–3 and has a height of 0.68 m. The stripping section is composed of stages 4–5 and has a height of 1.195 m.

Thermex is the name of the hot utility used in reboilers. The fluid is the Dowtherm™. A saturated vapor at 280°C, as the design specified. From the fluid data sheet, the latent heat, and the overall heat transfer coefficient were obtained. The Thermex fluid parameter values are given in **Table 6**.

According to design data, Thermex incoming and outgoing temperatures in the reboilers are equal, and this means that the Thermex vapor is saturated, and only latent is rejected from the hot utility.

Table 7 lists the input data for Column D-1 feed stream in the process simulation for the simulation model evaluation.

For the first simulation scenario, Column D-1 and D-2 bottom temperatures were set to 256°C and 277°C, respectively. The reboilers' heat loads were varied so the required separation specification was met. Though the heat load values were kept close to the plant design value. The validation results are given in **Table 8**.

Overall, the simulator was able to reproduce with good accuracy the actual conditions, although some discrepancies can be observed. Column D-1 bottom presented a very good correlation between simulated and actual data for both major components (nonylphenol and dinonylphenol). However, the simulation yielded a relatively higher phenol content at the bottom compared to the actual data. This occurs as the phenol quantity in the bottom stream is small in absolute numbers both in the simulated results and the actual date. Therefore, small deviations can result in high relative

Parameter	Value
Flow rate (kg/h)	4059
Temperature (°C)	156
Nonylphenol, %p	52.551
Dinonylphenol, %p	2.434
Phenol, %p	39.838
Nonene, %p	5.177

Table 7.
Feed stream input data according to design data.

Parameter	Design	Simulation	Variation
D-2 feed stream (kg/h)	2231.8	2244.4	0.56%
D-1 top temperature (°C)	138	134	−0.97%*
D-1 bottom temperature (°C)	256	256	0.00%*
E-4 heat load (kcal/h)	343,000	361,489	5.39%
E-4 Thermex flow rate (kg/h)	5170	5321	2.92%
D-2 feed—nonylphenol, %p	95.404	95.046	−0.38%
D-2 feed—dinonylphenol, %p	4.427	4.402	−0.56%
D-2 feed—phenol, %p	0.168	0.552	228.37%
D-2 feed—nonene, %p	0.001	0.000	−74.54%
D-2 distillate flow rate (kg/h)	2164.5	2131.05	−1.55%
D-2 bottom flow rate (kg/h)	67.2	113.35	68.67%
D-2 top temperature (°C)	215	209	−1.22%*
D-2 bottom temperature (°C)	277	277	−0.05%*
E-5 heat load (kcal/h)	146,000	158,418	8.51%
E-5 Thermex flow rate (kg/h)	2200	2332	5.99%
D-2 distillate—nonylphenol, %p	98.064	99.419	1.38%
D-2 distillate—dinonylphenol, %p	1.762	0.000	−99.99%
D-2 distillate—phenol, %p	0.173	0.581	235.84%
D-2 distillate—Nonene, %p	0.001	0.000	−73.19%

**temperature variations were calculated using the temperature values in Kelvin.*

Table 8.
Comparison between simulation results and design condition.

variation. As for Column D-1, both reboiler (E-4) heat load and Thermex flow rate present also low deviation values, 3% and 5% respectively.

For Column D-2 distillate stream, only nonylphenol and nonene presented low relative variation values, and the reason is similar to the one described for Column D-1 bottom stream. An important difference is in the dinonylphenol subproduct stream (bottom), where the simulation yielded a value 68% higher than the actual value. This difference can lead to false optimal points.

Properties	Jan/17	Jun/2018	Mar/2019
Flow rate (kg/h)	3535	3030	2713
Temperature (°C)	158.0	152.5	155.0
Nonylfenol, %p	51.134	49.317	48.318
Dinonylphenol, %p	1.502	1.641	1.762
Phenol, %p	35.143	30.283	30.706
Nonene, %p	12.220	18.759	19.214

Table 9.
Column D-1 feed stream properties.

For the evaluated production months, Column D-1 feed stream properties are given in **Table 9**.

Table 10 shows the comparison between simulation results using the D-1 feed stream average flow rate and composition for 2019, and the evaluated 3-month data.

For both the simulation results and the evaluated 3 months, Reactor R-2 effluent stream has a low dinonylphenol content. Hence, D-2 bottom stream (dinonylphenol subproduct) flow rate is small compared to the D-2 distillate stream (nonylphenol

Property	Average 2019	Jan/2017	Jun/2018	Mar/2019
Column D-2 feed stream flow rate (kg/h)	1438	1376	1332	1309
Column D-1 top temperature (°C)	134	131	127	127
Column D-1 bottom temperature (°C)	256	256	256	256
Reboiler E-4 heat load (kcal/h)	231,536	219,220	217,084	214,487
Reboiler E-4 hot utility flow rate (kg/h)	3571	3381	3348	3308
D-1 Bottom stream—nonylphenol, %p	95.046	96.627	96.258	95.957
D-1 Bottom stream—dinonylphenol, %p	4.402	2.835	3.201	3.499
D-1 Bottom stream—phenol, %p	0.552	0.538	0.540	0.543
D-1 Bottom stream—nonene, %p	0.000	0.001	0.001	0.001
D-2 Distillate stream flow rate (kg/h)	1364	1331	1284	1257
D-2 Bottom stream flow rate (kg/h)	73.82	44.71	48.62	52.37
Column D-2 top temperature (°C)	209	212	212	212
Column D-2 bottom temperature (°C)	277	277	278	278
Reboiler E-5 heat load (kcal/h)	101,467	99,733	96,331	94,417
Reboiler E-5 hot utility flow rate (kg/h)	1565	1538	1486	1456
D-2 distillate stream—nonylphenol, %p	99.419	99.444	99.438	99.433
D-2 distillate stream—dinonylphenol, %p	0.000	0.000	0.000	0.000
D-2 distillate stream—phenol, %p	0.581	0.556	0.560	0.565
D-2 distillate stream—nonene, %p	0.000	0.001	0.001	0.001

Table 10.
Comparison between simulation results and plant data.

product) flow rate. In comparison to **Table 3** data, **Table 10** shows a significantly lower dinonylphenol content in the D-2 distillate stream. A possible explanation for this difference is that the plant usually operates Column D-2 with a low reflux ratio to deliberately produce a high dinonylphenol content product stream but still in compliance with market demands.

Column D-2 temperature profile is not altered significantly with the flow rate. This is expected as feed composition has a much greater effect on column temperature than feed flow rate. The results prove that as composition varies, temperature varies significantly. Another parameter that has a direct influence on column temperature is the column operating pressure.

As the nonylphenol product becomes purer, the stream's flow rate decreases. The lower dinonylphenol content in R-2 effluent makes the separation in Column D-2 easier and that can be noted in the lower reflux ratio. Hence, this reduces the plant's operating cost.

Since composition has a larger effect on column performance than feed flow rate, simulation results are obtained for different dinonylphenol to nonylphenol ratios (DNF). The DNF ratios for the 3 evaluated months are given in **Table 11**.

In January 2017, the largest DNF ratio variation occurred as the dinonylphenol content in D-2 feed stream reaches the lowest value. This means that for such composition, the D-1 distillate stream already meets market requirements, and this stream could even completely by-pass Column D-2.

3.7 Energy consumption analysis

As columns' operating conditions are altered, an assessment of the reboilers' heat loads is made. These heat loads have a direct influence on the hot utility heat load, and also on the financial expenditure to operate the reboilers. **Table 12** shows the heat load values per nonylphenol produced.

For Reboiler E-4, only in March 2019, the heat load surpassed the 2019 average values. For Reboiler E-5, the heat load is lower than the average 2019 value for all months. The same is repeated for the sum of both heat loads.

Period	Jan/2017	Jun/2018	Mar/2019
DNF ratio	0.034	0.038	0.042
Variation from 2019 average	−36.18%	−28.02%	−20.83%

Table 11.
DNF ratio variation for the 3 evaluated months.

Reboiler	2019 average	Jan/2017	Jun/2018	Mar/2019
E-4 (D-1) (kJ/kg NF)	710.37	690.07	708.65	721.40
E-5 (D-2) (kJ/kg NF)	316.01	313.94	314.46	314.19
Total (kJ/kg NF)	1026.38	1004.01	1023.11	1035.59
Variation from 2019 average		−2.18%	−0.32%	−0.90%

Table 12.
Simulation results for reboiler heat loads.

	Jan/2017	Jun/2018	Mar/2019
Difference (R$)	−827.27	−120.98	−340.63

Table 13.
Expenditure variation for the heat loads' sum in the 3 evaluated months.

Therefore, by repeating January 2017 conditions, a 2% energy reduction is attained.

Thermex vapor is generated in a natural gas boiler. The following considerations are made: 100% efficiency in heat transfer from natural gas combustion to Thermex, the natural gas inferior calorific value of 33,500 kJ/m^3, the natural gas density is 0.7902 kg/m^3 (IEA, 2019), and the natural gas cost is R$2.48/$m^3$. Therefore, the heat load cost is $6.29.10^{-5}$ R$/kJ. A financial assessment is made according to the mentioned values and **Table 12** values. The results are given in **Table 13**.

Using the January 2017 value, the expenditure savings in a whole year could amount to R$ 9927.19.

3.8 Minimization of dinonylphenol production

Based on the months when dinonylphenol subproduct stream flow rate was minimal, the operating parameters of the nonylphenol production unit during these months are recommended as operating setpoints. As January 2017 presented the lowest dinonylphenol subproduct stream flow rate, the month's parameters are given in **Table 14**.

3.9 Analysis of the operating pressure effect

As the purity of distillation column outgoing streams is defined according to process requirements, the distillation column design depends on the definition of two degrees of freedom: the column operating pressure and the reflux ratio (or the number of stages). Column operating pressure influences the equilibrium curve for the heavy and light key components, the column temperature, the condenser and reboiler heat loads, the condenser and reboiler exchange areas, the required utilities, the

Parameter	Values
D-1 feed flow rate (kg/h)	2600
Reactor conversion	0.44
Nonene/raw phenol ratio	1.36
Nonene/recycled phenol ratio	0.588
D-1 reflux ratio	0.1
D-2 reflux ratio	0.1
D-1 operating pressure (bar abs)	0.333
D-2 operating pressure (bar abs)	0.0733
DNF/NF stream flow rate ratio	0.0285

Table 14.
Recommended new operating setpoints.

column diameter, and the column wall thickness, among other factors [12]. Therefore, defining the pressure is a crucial step in distillation column design.

The greatest challenge in defining the column operating pressure is that there is no established methodology for doing so, instead a handful of heuristics are often used and that can lead to sub-optimal designs [13]. Thorough optimization studies can be performed to attain a more cost-efficient design [14] but such studies are not always feasible as many economical parameters can be at the maximum, estimated with a reasonable degree of certainty. Even if the parameters were known, optimizing a distillation column is a time-consuming effort, and depending on the software and hardware used, the obtained solution is likely to be a local optimal and not a global optimal. That does not mean that operating pressure optimization should not be pursued but it means that engineers prefer to use heuristics to quickly attain a solution regardless of whether it is an optimal one or a feasible one.

In spite of optimization study difficulties, project and process engineers must comprehend the operating pressure effects on the distillation column operation. Firstly, relative volatility between light and heavy key components is usually inversely proportional to the operating pressure [15]. Hence, the lower the column operating pressure is, the higher the relative volatility is, and this means that at lower pressure, separation between key components can be made in fewer stages and with a lower reflux ratio. Evidently, how much pressure affects relative volatility depends on the components being separated. Using Column D-1 feed stream data from **Table** 7, the minimum number of stages, minimum reflux ratio, and temperature profiles for Columns D-1 and D-2 are analyzed as a function of the column operating pressure. The results for pressure variation are given in **Tables 15** and **16**. For **Table 16**, the simulations are made for a D-1 operating pressure of 0.240 bar abs.

As nonylphenol and dinonylphenol have a considerable difference in molecular weight, and consequently in relative volatility, the results in **Tables 15** and **16** show that pressure variation has a small effect on the minimum number of stages and minimum reflux ratio. This small effect is also due to the relatively narrow simulated pressure variation range. As for the columns' bottom and top temperatures, despite the very low pressures, the temperatures are high. This is expected since dinonylphenol is a large molecule with a very high boiling point. As a suggestion for the plant's future optimization efforts, operating Column D-1 with an even lower

Operating pressure (bar)	Minimum number of stages	Minimum reflux ratio	Bottom temperature (°C)	Top temperature (°C)
0.213	3.28	0.0129	257	114
0.227	3.30	0.0134	259	116
0.240	3.32	0.0138	261	118
0.253	3.34	0.0142	263	119
0.267	3.37	0.0146	265	121
0.280	3.39	0.0150	266	122
0.293	3.41	0.0154	268	124

Table 15.
Column D-1 operating pressure variation effect.

Operating pressure (bar)	Minimum number of stages	Minimum reflux ratio	Bottom temperature (°C)	Top temperature (°C)
0.0600	3.08	0.0114	318	206
0.0633	3.10	0.0117	320	207
0.0667	3.11	0.0120	322	209
0.0700	3.13	0.0123	324	210
0.0733	3.14	0.0126	325	212
0.0767	3.15	0.0128	327	213
0.0800	3.17	0.0131	329	214

Table 16.
Column D-2 operating pressure variation effect.

pressure could yield a bottom temperature where Thermex could be replaced with high-pressure steam. As for Column D-2, this is not possible, as this column already operates near full vacuum.

4. Conclusions

The study aimed at subproduct (dinonylphenol) reduction and also heat load reduction. For these aims, a thorough investigation of the design conditions, and historical plant data was made to determine the months that yielded the lowest dinonylphenol production. The plant relies on two strategies to reduce the dinonylphenol production, and these are defining reactor conditions that minimize conversion of nonylphenol into dinonylphenol and increasing dinonylphenol content in the product (nonylphenol) stream to the maximum allowable composition. Aspen Plus® was the simulator used to investigate a new operating condition for the nonylphenol production unit to attain the study goals.

The results indicate that a 36% subproduct reduction is possible. This value is achieved by a comparative analysis between simulation results and data from months where subproduct production reached minimum values. By simulating the nonylphenol plant design conditions, operating parameters are obtained for the reaction, phenol recovery, and nonylphenol purification stages. These recommended parameters are validated using historical plant data. The new simulated condition reduces the reboilers' heat loads, and this can lead to a 2% reduction in hot utility consumption.

Author details

Julio Cesar Ribeiro Nunes[1], Ricardo de Freitas Fernandes Pontes[1],
Fabio Rodolfo Miguel Batista[2] and Rafael M. Matricarde Falleiro[1*]

1 Department of Chemical Engineering, Federal University of São Paulo (UNIFESP), Diadema, SP, Brazil

2 Department of Chemical Engineering (DEQUI-EEL), Lorena Engineering School, University of Sao Paulo (USP), Lorena, SP, Brazil

*Address all correspondence to: falleiro@unifesp.br

References

[1] Lorenc JF, Lambeth G, Scheffer W. Alkylphenols. In: Kirk-Othmer Encyclopedia of Chemical Technology. New York: John Wiley & Sons, Inc; 2003. DOI: 10.1002/0471238961.0112112512151805.a01.pub2

[2] Arné M. Nonionic Surfactants. California: SRI International: Process Economics Program; 1984

[3] Dow Chemical Company. Dowtherm™ A Heat Transfer Fluid: Product Technical Data. 2011. [Online]. Available from: https://www.appliedthermalfluids.com/wp-content/uploads/2018/02/Dowtherm-A-heat-transfer-fluid-MSDS.pdf

[4] Olujić Ž. Chapter 1—Types of distillation column internals. In: Górak A, Olujić Ž, editors. Distillation Equipment and Processes. Cambridge, Massachusetts, USA: Academic Press; 2014. pp. 1-34. DOI: 10.1016/B978-0-12-386878-7.00001-2

[5] McCabe WL, Smith JC, Harriot P. Unit Operations of Chemical Engineering. 7th ed. New York, NY, USA: McGraw Hill Chemical Engineering Series; 2007

[6] Stewart MI. Condensate stabilization. Surface Production Operations (Gulf Professional Publishing). 2014;**2**:259-277. DOI: 10.1016/b978-0-12-382207-9.00006-8

[7] Lange A, Fieg G. Designing novel structured packings by topology optimization and additive manufacturing. Computer Aided Chemical Engineering. 2022;**49**: 1291-1296. DOI: 10.1016/B978-0-323-85159-6.50215-3

[8] Dzhonova-Atanasova D, Stefanova K, Nakov S. On liquid flow maldistribution through investigation of random open-structure packings. Designs. 2023;7:47. DOI: 10.3390/designs7020047

[9] Jerzy M. Prediction of separation efficiency of structured and stacked packings under low and normal pressure. Chemical Engineering Research and Design. 2022;**186**:713-729. DOI: 10.1016/j.cherd.2022.07.016

[10] Sandler SI. Using Aspen Plus® in Thermodynamics Instruction. New Jersey: John Wiley & Sons, Inc; 2015

[11] Al-Malah KIM. Aspen plus®: Chemical Engineering Applications. New Jersey: John Wiley & Sons Inc; 2017

[12] Liu ZY, Jobson M. The effect of operating pressure on distillation column throughput. Computers and Chemical Engineering Supplement. 1999;**23**:S831-S834. DOI: 10.1016/S0098-1354(99)80204-X

[13] Luyben WL. Distillation column pressure selection. Separation and Purification Technology. 2016;**168**: 62-67. DOI: 10.1016/j.seppur.2016.05.015

[14] Cui C, Liu S, Sun J. Optimal selection of operating pressure for distillation columns. Chemical Engineering Research and Design. 2018;**137**:291-307. DOI: 10.1016/j.cherd.2018.07.028

[15] Wankat PC. Separation Process Engineering—Includes Mass Transfer Analysis. 3rd ed. New Jersey, NY, USA: Prentice-Hall; 2012

Chapter 7

Membrane Distillation Process: Fundamentals, Applications, and Challenges

Ali Boubakri, Salah Al-Tahar Bouguecha and Amor Hafiane

Abstract

Traditional thermal-based processes such as multistage flash and multi-effect distillation have been used for thousands of years to obtain freshwater from saline water. Recently, with the development of membrane-based technology, membrane distillation (MD) as a thermally driven membrane process has received significant attention. The driving force in MD is the vapor pressure gradient induced by temperature difference through hydrophobic microporous membrane pores. The membrane used for MD should be hydrophobic and microporous. In MD, the mechanism of transport involves simultaneously heat and mass transfers, which moves from the hot feed side to the cold permeate side. The performance of MD is evaluated based on various performance metrics including permeate flux, recovery ratio, thermal efficiency, gained output ratio, and specific thermal energy consumption. It has good ability for various industrial uses due to its moderate applied temperature and pressure, high rejection rate, less membrane fouling tendency and its ability to treat high-saline water. The water production cost still remains high compared to conventional processes. Therefore, MD can be cost-effectively when integrated with solar energy, geothermal energy and waste heat. Nevertheless, MD process requires focused research to improve its efficiency to become more mature and economically competitive at large scale.

Keywords: membrane distillation, membrane, Transport models, hybrid systems, desalination

1. Introduction

Membrane-based technologies are becoming more attractive for many industrial separation processes, especially in water treatment sector due to their various benefits over thermal-based technologies. The advantages include cost and thermal energy efficiencies, better environmental impacts with less fuel combustion needed to generate thermal energy, and lower pumping and capital costs [1]. Recently, the growing demand for freshwater water has been at high level, explained by the rise of the global water desalination market which was valued at US $15.43 billion in 2017, expected to reach US $27 billion by 2025 [2].

The global water reserve available as freshwater is estimated to be 3%, the remaining 97% is available as seawater. In the coming decades, climate change and population growth are expected to exacerbate the availability of freshwater sources due to shifting rainfall pattern, droughts, temperature pattern, and anthropogenic factors. Non-conventional sources of water, such as desalinated water and recycled/reused wastewater, are considered as the most viable option aimed to alleviate the intensity of water shortage under circular water economy approach. The main membrane-based technological options applied for water desalination and water treatment are microfiltration (MF), ultrafiltration (UF), nanofiltration (NF), membrane bioreactor (MBR), reverse osmosis (RO). However, conventional membrane-based technologies have several limitations including high energy demand, high cost of operating and maintenance, and membrane performance reducing due to fouling phenomena. Conventional thermal-based technologies have been limited by its high need of thermal energy and high cost of operating and maintenance.

Membrane distillation (MD) is an emerging technology that has a consistent margin progression compared to conventional commercialized technologies such as RO, NF, MED, and MSF and has been explored in the field of water desalination and wastewater treatment. MD integrating both thermal distillation and membrane processes. MD is a phase change process, where the driving force is the difference between vapor pressures, induced by temperatures gradient, between two solutions separated by hydrophobic microporous membrane. The main benefit of MD is the ability to operate at lower operating temperature, below its boiling point, compared to conventional thermal-based technologies and at lower hydrostatic pressure compared to conventional pressure-driven membrane technologies. More recently, there has been growing interest in MD technology for its potential to treat high-saline streams with low brine discharge and its aptitude of integration with other membrane-based processes and low-grade heat energies.

Interest in the technology of MD was first proposed by Bodell [3] in 1963, who first explained the process and described the approach to produce potable water. Later, in 1967, Findley [4] published the first documents with interesting results related to MD process. From 1970s to 1990s, MD has gone through long development phase with slow growth. During the last 20 years, there has been a rapid growth with a considerably increase of the number of scientific publications on MD technology, as displayed in **Figure 1**. Despite the enhanced research interest within the academic community, MD technology is not widely commercialized yet. The main reasons are the lack of suitable designed modules and adequate and the high thermal energy cost.

In this chapter, fundamental knowledge, membranes and basic configurations, transfer model equations, performance metrics for main applications, and possible integration of the MD system are presented. The understanding of the fundamental mechanisms of MD process is crucial for researchers and engineers in the field of water desalination and water treatment to apply and analyze MD system more effectively. In addition, this chapter provides comprehensive and valuable information on the evaluation of MD performance, which will be a useful guide to studying MD technology.

2. Fundamentals of MD

2.1 Principles of MD

MD is a thermally-driven membrane separation process, only vapor molecules transfer across a microporous hydrophobic membrane [5]. The driving force during

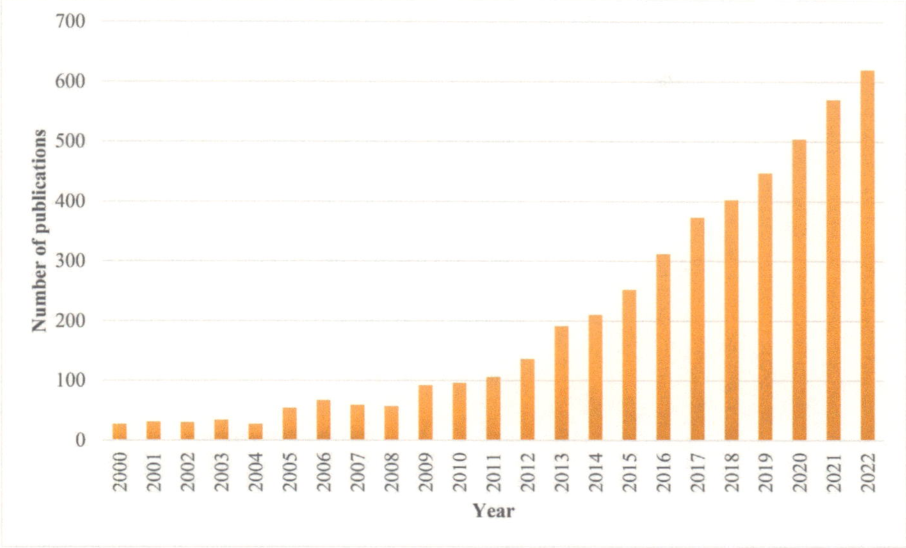

Figure 1.
Evolution of number of research publications related to MD technology during the period 2000–2022.

MD process is the difference between vapor pressures between two membrane sides generated by a temperature gradient by circulating the hot feed solution (FS) and the cold permeate solution. The vapor generated on the hot FS passes across the membrane and condenses on the cold solution to generate the distillate. The vapor pressure difference increases with the increase of temperature difference between feed and permeate solutions. The main advantage of MD is that considered as low temperature process, it can be operated typically at temperature lower than 80°C [6]. **Figure 2** displays the principle of MD process using for the desalination of seawater to produce desalinated freshwater. In the MD process, the hydrophobic nature of the used membrane allows water vapor molecules in the hot feed to pass through the pores of the membrane, while prevents the penetration of the liquid feed water into the membrane pores due to its low surface energy [7]. As a result, a liquid–vapor interfaces are formed at the entrances of the hydrophobic membrane pores. Basically, the pores of the membrane used for MD must be kept dry to prevent the liquid feed water to pass through the hydrophobic membrane.

2.2 Basic MD configurations

In this section, different MD configurations that have been utilized to separate feed solution (FS) using a microporous hydrophobic membrane will be presented. **Figure 3** shows the four basic configurations of MD process.

2.2.1 Direct contact membrane distillation

Direct contact membrane distillation (DCMD) is the simplest and the most studied configuration in MD. In this configuration, the hot FS is in direct contact with one side of the membrane and the cold permeate solution is in direct contact with the other

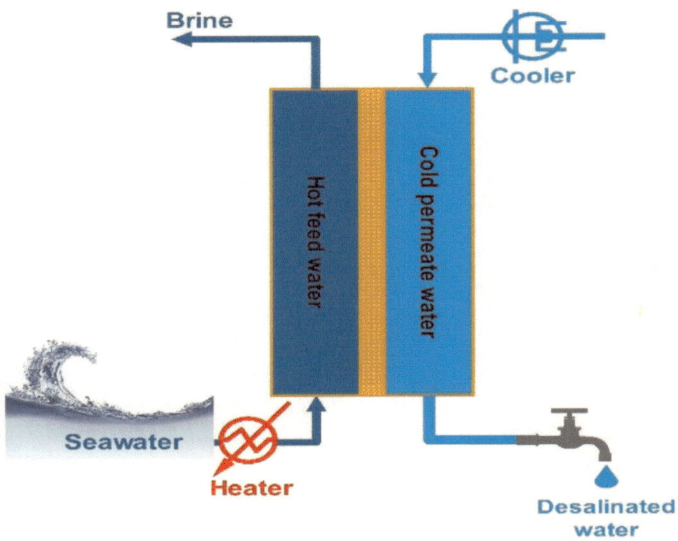

Figure 2.
The principle of MD.

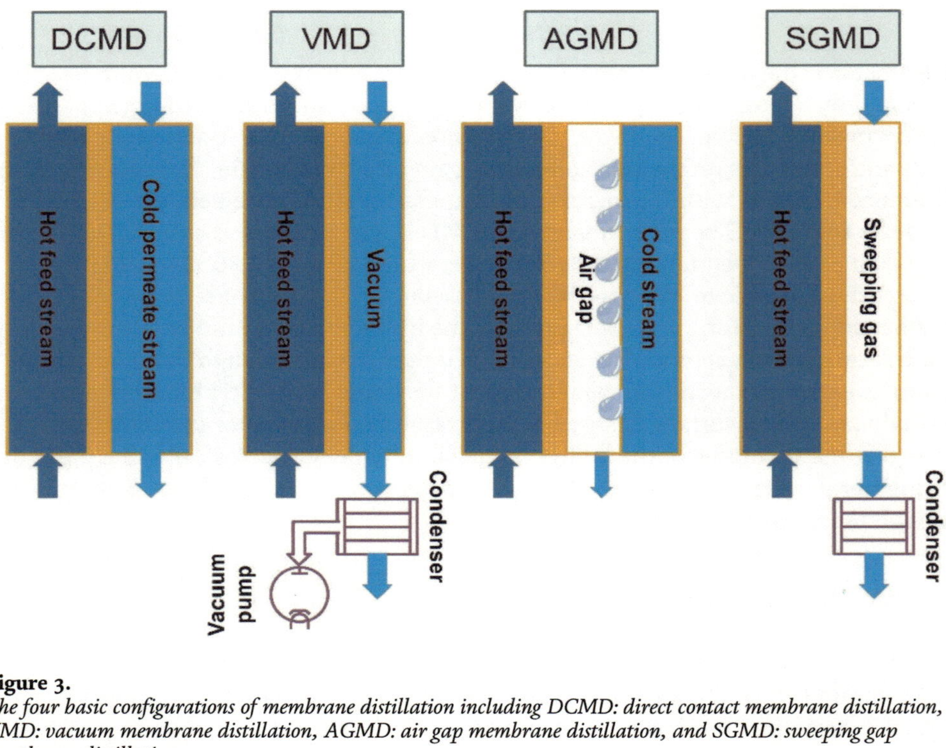

Figure 3.
The four basic configurations of membrane distillation including DCMD: direct contact membrane distillation, VMD: vacuum membrane distillation, AGMD: air gap membrane distillation, and SGMD: sweeping gap membrane distillation.

side of the membrane (**Figure 3**—DCMD). Therefore, solution evaporation takes place at the feed-membrane side and transferred across the membrane porous to condense on the other side in the liquid permeate solution. The hydrophobic characteristic of the membrane prevents the liquid solution to penetrate the membrane porous, only vapor solution can penetrate. The main drawback of the DCMD configuration is its comparatively large conductive heat losses owing to the contact between hot feed side, membrane, and permeate cold side [8].

2.2.2 Vacuum membrane distillation

The schematic diagram of the vacuum membrane distillation (VMD) configuration is presented in **Figure 3**—VMD. In this configuration, the vapors created in the permeate side are sucked by a vacuum created in the cold side by means of vacuum pump. The condensation takes place outside the module. The applied vacuum pressure should be lower than the saturation pressure created in the hot FS in order to provide sufficient driving force [7]. The advantages of VMD are its high transmembrane water flux and small conductive heat losses. Despite that, VMD presents higher membrane fouling propensity and lower pore-wetting resistance compared to other configurations.

2.2.3 Air gap membrane distillation

In air gap membrane distillation (AGMD), as the schematic diagram shown in **Figure 3**—AGMD, a stagnant air is introduced between the permeate-membrane side and the condensation surface. The evaporated molecules coming from the hot FS cross both the membrane pores and the air gap to condensate at the cold surface inside the MD module. AGMD shows low heat lost by conduction, which leads to high energy efficiency, but the obtained water flux is generally low due to the additional resistance to mass transfer created by the air gap.

2.2.4 Sweeping gas membrane distillation

The schematic diagram of the VMD configuration is presented in **Figure 3**—SGMD. In sweeping gas membrane distillation (SGMD), a cold inert gas is applied to sweep the vapor molecules at the permeate-membrane side which is then condensate outside the MD module by means of an external condenser. In this configuration, the sweep gas reduces the mass transfer resistance and consequently enhances the mass transfer coefficient. However, it requires a large sweep gas volume, leading to an increase in condenser capacity. This, in turn, increases additional cost and the system complexity.

2.3 Membrane characteristics

The MD technology is a membrane-based system. The characteristics of the used membranes are of great importance to determine the performance of the process. In MD, the used membranes should be hydrophobic (non-wetting) microporous. Generally, these membranes are made from polyvinylidene fluoride (PVDF), polytetrafluoroethylene (PTFE), or polypropylene (PP) materials. A suitable membrane for MD should have high hydrophobicity, high porosity, uniform pore size distribution, high liquid entry pressure (LEP), low thermal conductivity, high resistance to chemicals, excellent mechanical strength, and low fouling tendency.

Hydrophobic membrane materials should be used to reduce its wettability. The hydrophobicity is evaluated by the measurement of the surface contact angle. The membrane porosity determines the ratio between the volume of the pores and the total volume of the membrane. Higher membrane porosity can guarantee lower conductive heat loss and higher permeate flux. In general, the pore size ranged between 0.1 and 1 μm and the porosity varied between 40% and 90% [8]. The LEP is the minimum hydraulic pressure that must be applied to a liquid before it overcomes the hydrophobicity of the membrane and penetrating the pores [9]. To achieve high MD performance, the value of LEP is suggested to be more than 2.5 bar. A low thermal conductivity is crucial to minimize the heat losses through conduction, hence maintain the driving force for the movement of water vapor across the membrane. Generally, the used materials have thermal conductivity ranged between 0.1 and 0.5 W m^{-1} K^{-1}. Additionally, high resistance to chemicals (e.g., acids and bases), excellent mechanical strength, and low fouling tendency are crucial characteristics of membranes to ensure MD process performance stability during long-term operation.

2.4 Transport models

2.4.1 Heat transfer

In MD, the mechanism of transport involves simultaneously heat and mass transfers, which occur in the same direction from the hot feed side to the cold permeate side. The heat transfer is carried out in four steps: (1) the heat transferred from the bulk FS to the feed-membrane interface; (2) the heat transferred by conduction through the microporous membrane; (3) the heat associated to the latent heat of vaporization through the membrane pores; and (4) the heat transferred from the permeate-membrane interface to the bulk permeate solution [10, 11].

The heat transferred from the bulk FS to the feed-membrane interface:

$$Q_f = h_f(T_f - T_{fm}) \tag{1}$$

The heat transferred by conduction through the membrane associated with the heat associated to the latent heat of vaporization through the membrane pores:

$$Q_m = Q_v + Q_c = J\Delta H_v + h_m(T_{fm} - T_{pm}) \tag{2}$$

The heat transferred from the permeate-membrane interface to the bulk permeate solution:

$$Q_p = h_p(T_p - T_{pm}) \tag{3}$$

Where h_f, h_p, and h_m are the heat transfer coefficient in feed side, permeate side, and membrane respectively. T_f and T_p are the bulk feed and bulk permeate temperatures. T_{fm} and T_{pm} are the temperature at the feed-membrane and permeate-membrane interfaces, respectively. Q_v and Q_c are the heat energy of vaporization and heat of condensation, respectively. J is the mass transfer of vapor through the membrane porous. ΔH_v is the latent heat of vaporization.

The overall heat transfer flux through the whole MD system is given by the following equation:

$$Q = Q_f = Q_p = Q_m \quad (4)$$

2.4.2 *Mass transfer*

The mass transport in MD involves three major stages that occur in series: (1) vapor generation takes place at the feed-membrane interface; (2) vapor transport across the membrane pores; and (3) vapor condensation at the permeate-membrane interface. The mass transfer of vapor through the membrane porous is generally defined according to Darcy's law, expressed as follows [12]:

$$J = B_m (P_{m,f} - P_{m,p}) \quad (5)$$

Where B_m is mass transfer coefficient, $P_{m,f}$ is the partial vapor pressure at the feed-membrane interface, and $P_{m,p}$ is the partial vapor pressure at the permeate-membrane interface.

According to the dusty gas model (DGM), the prediction of mass flux through a porous membrane can be controlled by three basic mass transfer mechanisms, including Knudsen diffusion, viscous or Poiseuille flow, molecular or ordinary diffusion [13]. To determine the dominant mass transfer mechanism of flow through the membrane pores, it is necessary to calculate the Knudsen number (K_n), from the following equation:

$$K_n = \frac{\lambda}{d_p} \quad (6)$$

Where d_p is the pore diameter and λ is the mean free path, which can be calculated from the following equation:

$$\lambda = \frac{k_B T_m}{\pi * \left(\frac{\sigma_w - \sigma_a}{2}\right)^2 P \sqrt{1 + \left(\frac{M_w}{M_a}\right)}} \quad (7)$$

Where k_B is the Boltzmann constant. T_m is the average temperature of the membrane. σ_w and σ_a are the collision diameters of water and air; respectively. P is the total pressure inside the pore. M_w and M_a are the molecular weights of the water and the air, respectively.

If the Knudsen number is less than 0.01, the mass transfer mechanism is dominated by the molecular diffusion. If the Knudsen number is ranged between 0.01 and 1, the mass transfer mechanism is dominated by the combined Knudsen-molecular diffusion. If the Knudsen number is larger than 1, the mass transfer mechanism is dominated by the Knudsen diffusion [14].

Therefore, the flux of vapor through the membrane porous can be calculated using the following equations for molecular, Knudsen-molecular, and Knudsen diffusion, respectively:

For molecular diffusion:

$$J = \frac{\varepsilon * D_w * M_w}{\tau * \delta * P_a * R * T} (P_{m,f} - P_{m,p}) \quad (8)$$

For Knudsen-molecular diffusion:

$$J = \left[\frac{3 * \tau * \delta}{\varepsilon * d_{\mathrm{p}}} \sqrt{\frac{\pi * R * T}{8 * M_{\mathrm{w}}}} + \frac{\tau * \delta * P_{\mathrm{a}} * R * T}{\varepsilon * PD_{\mathrm{w}} * M_{\mathrm{w}}}\right]^{-1} \left(P_{\mathrm{m,f}} - P_{\mathrm{m,p}}\right) \quad (9)$$

For Knudsen diffusion:

$$J = \frac{\varepsilon * d_{\mathrm{p}}}{3 * \tau * \delta} \sqrt{\frac{8 * M_{\mathrm{w}}}{\pi * R * T}} \left(P_{\mathrm{m,f}} - P_{\mathrm{m,p}}\right) \quad (10)$$

Where ε is the membrane porosity, τ is the membrane tortuosity, δ is the membrane thickness, D_{w} is the diffusion coefficient of water, P_{a} is pressure of air, P is the total pressure inside the pore, R is the gas constant, and T is the temperature.

3. Membrane distillation metrics

Some performance indicators have been used to evaluate the efficiency of MD process such as permeate flux (J_{p}), recovery rate (Y), thermal efficiency (η), gained output ratio (GOR), and specific thermal energy consumption (STEC).

3.1 Permeate flux

Permeate flux is the main parameter used as performance indicator for MD process, it presents the productivity of the system. The permeate flux can be defined as the water flow rate passing through a specified membrane area at a certain time. It is calculated using the following equation:

$$J_p = \frac{m_p}{S * \Delta t} \quad (11)$$

Where m_{p} is the produced water weight, S is the effective membrane surface, and ΔT is the time interval.

3.2 Recovery ratio

The recovery ratio is a significant performance criterion which is used to determine the design size and the economic aspect of MDs system. It is defined as the percentage of the produced permeate flow rate over the feed flow rate.

$$Y = \frac{m_{\mathrm{p}}}{m_{\mathrm{f}}} * 100 \quad (12)$$

Where m_{f} is the feed water weight.

3.3 Thermal efficiency

In MD, the heat transfer is divided into two steps including heat transfer through membrane matrix by conduction and heat transfer by movement of vapor across the membrane induced by latent heat of vaporization. The heat transfer through membrane matrix by conduction is considered as heat loss, which must be minimized to improve the thermal efficiency of the membrane. Thermal efficiency is defined as the ratio of the useful energy delivered (latent heat of vaporization) to the total heat (latent and conduction) and is calculated using equation as follows:

$$\eta = \frac{J_p \Delta H_v}{J_p \Delta H_v + \left(\frac{k_m}{\delta}\right)\left(T_{fm} - T_{pm}\right)} * 100 \tag{13}$$

Where ΔH_v is the latent heat of vaporization, T_{fm} and T_{pm} are the temperature at the feed-membrane and permeate-membrane interfaces, respectively, k_m is the thermal conductivity of the membrane, δ is the membrane thickness.

3.4 Gained output ratio

The GOR is used to assess the energy efficiency of MD system. GOR can be defined as the ratio of the thermal energy required to vaporize the mass of water produced to the total thermal energy provided to the system [15]. GOR is calculated using the following equation:

$$\text{GOR} = \frac{F_d\, S\, \rho_d\, \Delta H_v}{F_f\, \rho_f\, C_{p,f}\, \left(T_{f,in} - T_{p,out}\right)} \tag{14}$$

Where F_d and F_f are the distillate water and feed water flow rates, respectively. ρ_d and ρ_f are the density of distillate and feed waters, respectively. $T_{f,in}$ and $T_{p,out}$ are the feed inlet and the permeate outlet temperatures. ΔH_v is the latent heat of vaporization. $C_{p,f}$ is the specific heat of feed water. S is the effective membrane surface.

3.5 Specific thermal energy consumption

The STEC is another thermal performance indicator for MD process. STEC represents the amount of external thermal energy consumed by MD system to produce a unit volume of distilled water and is calculated using the following equation:

$$STEC = \frac{F_f * \rho_f * C_{p,f} * \left(T_{f,in} - T_{p,out}\right)}{3.6 * 10^6 * F_d} \tag{15}$$

4. Potential applications

The main advantages of MD are its ability to achieve high rejection rate (theoretically 100%), and it operates at atmospheric pressure and relatively low temperatures. These supremacies of MD compared to conventional membrane-based technologies, render it a promising technology which gained significant attention for various potential applications.

MD can be applied for the desalination of saline water including seawater and brackish water. El Mokhtar et al. [16] investigated the feasibility of AGMD for the desalination of seawater. At feed temperature of 77°C, AGMD process produced freshwater with high salt rejection, more than 99%, and permeate flux of 9.06 kg m^{-2} h^{-1}. The comparison between experimental and predicted models shown high accuracy. Another study conducted by Usman et al. [17] provided a systematic evaluation of the economics involved in desalination of brackish water using MD process to produce potable water. The combination of solar-thermal sources with waste heat to power MD system greatly lowered the water production cost. MD system relying on waste heat and solar thermal can lower water production cost from \$6.80/m^3 to a mere \$1.6/m^3. Additionally, MD is less sensitive to high salinity, it can be used for the desalination of hypersaline solutions such as brine and concentrated wastewater toward near-zero liquid discharge.

Furthermore, high removal efficiency of small molecule contaminants and heavy metal ions is achieved by MD. Lou et al. [18] investigated the efficiency of MD process integrated with crystallization technology for the treatment of concentrated heavy metal wastewater, including zinc and nickel. Results show that MD process integrated with crystallization proves to be a promising technology for treating highly concentrated heavy metal solutions and the used membrane has excellent resistance to fouling following the treatment of highly concentrated solutions.

MD also has the potential to minimize the amount of wastewater discharged from various wastewater sources, such as municipal wastewater, textile wastewater, pharmaceutic industry wastewater, liquid nuclear wastewater, oily wastewaters, etc. Produced water, is a by-product wastewater effluent with high salinity, contains suspended particles, dispersed oils, and chemicals. MD can be applied for the treatment of highly saline produced water. MD also has the potential to recover valuable components. It can be utilized for the concentration of fruit juices, herbal extracts, sugars, alcohols, mineral acids, etc. Criscuoli and Drioli [19] evaluated the performance of VMD process to concentrate date juices at low temperature. After previous treatment steps (enzymatic treatment, clarification, decolorization, and deionization), the date juices can be processed by VMD process with low membrane fouling tendency.

5. MD-based hybrid processes

5.1 Integration with membrane-based processes

5.1.1 RO-MD

Pressure-driven RO is highly efficient desalination technique for production of freshwater. However, it is limited by its low capability to treat high concentrated streams due to physical limitations caused by high osmotic pressure values. The RO water recovery for the desalination of seawater is limited to 60% when using a double stage and applying high pressure. In this case, MD is integrated to treat the RO brine to overcome the limitation of water recovery. Hence, RO-MD integrated system can increase the overall water recovery and therefore reduce the brine discharge to achieve minimal liquid discharge approach. However, the main drawback of coupling MD with RO is the high MD membrane scaling propensity induced by the presence of high chemicals content in RO brine (salts, antiscalants, chemical cleaning, etc.)

5.1.2 FO-MD

Forward osmosis (FO) is an osmotically-driven separation process that utilizes the osmotic pressure difference between a FS and a draw solution (DS) to transport water molecules across a semi-permeable membrane [20]. It is considered as the lower required energy consumption among the desalination processes. The main drawback of FO is that it is limited by the dilution of DS, thus a recovery process must be added [5]. For this purpose, MD can be integrated with FO process to provide a good alternative for the regeneration and the reuse of DS as well as the generation of high-water quality. In FO-MD hybrid process, the FS is concentrated by the FO system, and the diluted DS is concentrated by the MD process to produce high osmotic pressure solution. The FO-MD integrated system can be applied for water desalination and wastewater treatment with high-water recovery and shown great potential in achieving zero liquid discharge approach.

5.1.3 MBR-MD

MBRs combine conventional bioreactor with pressure-driven UF or MF membranes under aerobic and anaerobic conditions. MBR is mainly used to separate suspended solids from domestic and industrial wastewater due to its simplicity, reliability, and cost-effectiveness. However, the used membranes are inefficient in retaining inorganic salts and trace organic contaminants, thus the process cannot attain water reclamation and reuse [21]. Therefore, MD can be integrated with MBR to form the hybrid membrane distillation bioreactors (MDBR) for efficient wastewater treatment and recovery. MDBR integrated system can be potentially used for water, nutrients, bioenergy, and recovery of value-added products from domestic and industrial wastewaters.

5.2 Integration with renewable energy

5.2.1 Solar energy

The lower operating temperatures needed for MD technology compared to conventional thermal processes make it possible to be coupled with alternative low-grade heat sources such as solar energy, geothermal energy, and waste heat. Using alternative energy sources can drastically lower the cost and energy consumption of the process, as well as reduce its environmental footprint.

Solar energy is widely used as renewable energy sources for thermal energy generation in MD. It used energy collected from sunlight to heat the FS for MD process. Collecting heat from solar energy can be generated by two methods: direct collecting and indirect collecting [22]. The direct collecting method consists of converting concentrated solar radiation into thermal energy through a moving fluid, as a kind of heat exchanger. The indirect collecting method consists of converting solar energy into electricity through photovoltaic cells. The generated electricity is then used directly or stored in batteries for heating the FS or operate various components in MD system such as pumps, compressors, etc. It can be reported that using solar energy-powered MD system instead of electricity-powered system, under the same conditions, can reduce the specific energy consumption by 30% and increase the GOR by 17% [23].

5.2.2 Geothermal energy

Unlike solar energy, geothermal energy is found in deep underground and does not depend on weather and climatic conditions. Geothermal energy sources are expected to reduce the energy cost and lower the overall water production cost in MD system.

The main advantage of using geothermal energy in MD is its stable heat source that is readily accessible, enabling the direct use for heating feed water without the need for an energy converter. The integration of low-grade heat sources of geothermal energy with MD is potentially applied for the desalination of geothermal brackish water and wastewater treatment. The scaling up potential of MD is related to its capability to use the steady heat source of geothermal energy which is better than using solar energy [10]. Despite the abundance of geothermal energy sources in many regions in the world, the integration of this kind of renewable energy is limited due to the limitation of adequate geographical sites.

5.2.3 Waste heat

Waste heat is a freely available low-grade energy source released at a temperature lower than 100°C from various industries such as power stations, diesel engines, nuclear reactors, and industrial plants, etc. As waste heat is a freely abundant energy source, it can be integrated with MD which improves the energy efficiency of the system with no additional carbon footprint. Thus, coupling MD with waste heat is a promising solution and becomes environmentally and economically competitive with conventional membrane-based technologies. The low-grade waste heat is integrated with MD used for the desalination of seawater and brackish water. It can be reported that MD powered by waste heat source reduces the energy cost to 0.31 USD/m^3, which is lower than those using conventional RO technology with 0.45 USD/m^3 [24].

6. Challenges in MD process

6.1 Polarization phenomena

6.1.1 Temperature polarization

In MD, the low thermal efficiency generated by temperature polarization is one of the key obstacles hampering its applications. The vaporization occurs at the membrane hot surface and the condensation occurs at the membrane cold surface, leading to the creation of a thermal boundary layer between two membrane sides. This effect means that a large amount of heat supplied to the process is transferred to the cold side by conduction instead of evaporating feed water. This phenomenon is called temperature polarization, estimated using the temperature polarization coefficient (TPC), which is defined as follows:

$$\text{TPC} = \frac{T_{\text{fm}} - T_{\text{pm}}}{T_{\text{f}} - T_{\text{p}}} \tag{16}$$

Where T_{fm} is the temperature on the feed-membrane side, T_{pm} is the temperature on the permeate-membrane side, T_{f} is the temperature of the bulk feed, and T_{f} is the temperature of the bulk permeate.

6.1.2 Concentration polarization

Like temperature polarization, concentration polarization takes place in MD process, but it has less effect on permeate flux compared to temperature polarization. MD process provides the theoretically 100% rejection of all non-volatiles, indicating that the concentration polarization only occurs in the feed side. An increase in concentration polarization led to an additional mass transfer resistance that will decline the permeate flux. The estimation of the concentration polarization is calculated by concentration polarization coefficient (CPC), which can be expressed as follows [25]:

$$\text{CPC} = \frac{C_{\text{fm}}}{C_{\text{f}}} \tag{17}$$

Where $C_{\text{f,m}}$ is the solute concentration on the feed-membrane side and $C_{\text{f,}}$ is the solute concentration in the bulk feed.

6.2 Membrane fouling

Membrane fouling is the accumulation of substances on the membrane surface or inside the membrane pores, which causes a deterioration of the overall performance of MD in terms of permeate flux and solute rejection. Generally, membrane fouling can be classified into three categories according to the source of feed water, including inorganic fouling or scaling, organic fouling, and biological fouling. Inorganic fouling is caused by the accumulation of salt precipitates such as calcium sulfate, calcium carbonate, silicate, and sodium chloride, etc. Organic fouling is caused by colloidal organic matters such as humic substances, extracellular polymeric substances, and proteins. Biological fouling (biofouling) is caused by the build-up of microorganisms on the membrane surface [26]. The main parameters that affect membrane fouling are the depositing materials characteristics (concentration, solubility, diffusivity, charge, etc.), and membrane structure (hydrophobicity, roughness, pore size, charge, etc.).

6.3 Membrane wetting

Membrane wetting is one of the main challenges faced in the development and commercialization of MD technology. Membrane wetting is a phenomenon in which the liquid FS freely permeates through the pores of the hydrophobic membrane, which causes a deterioration of membrane performance in terms of solute rejection. Membrane wetting is caused by the presence of surface-active materials, such as oils or surfactants which adsorb onto the hydrophobic MD membrane and make it progressively more hydrophilic [27]. In MD, the membrane-wetting phenomenon is detected simply by a simple measurement of distillate electrical conductivity.

7. Conclusion and future directions

MD is an emerging thermally-driven membrane-based system that has become one of the most attractive processes in sustainable techniques to help reduce the global water-energy stress in separation and water treatment applications. The integration of MD with conventional membrane-based processes such as RO, FO, and MBR, can significantly enhance its performance. Moreover, the integration of renewable and low-cost energy sources such as solar energy, geothermal energy, and waste heat can make MD technology cost- and energy-effective with more sustainable way. MD has the advantage to have higher solute rejection rate than pressure-driven technologies and lower thermal energy consumption than conventional distillation technologies.

MD also can treat high-saline waters with high recoveries generated near-zero liquid discharge approach applied for water desalination and wastewater purification. However, MD is still facing many challenges that hinder its wide industrial application, due to the relatively low water flux and high energy consumption conventional RO technology. Therefore, future directions require focused research to improve the efficiency of MD to become more competitive at large scale, among them:

1. Design and fabrication of appropriate membrane and modules for MD application. The purpose of this approach aimed to reduce the effect of temperature and concentration polarization in order to enhance the efficiency of the MD process. Membranes with high thermal conductivity, large pore size, high porosity, low tortuosity, small thickness are required to achieve better performance. Moreover, lack of suitable designed modules limits the deployment of MD technology at industrial scale. Exploring new module configurations suitable for the specificity of MD flows is also interesting.

2. Reduce the energy consumption for large-scale translation of MD technology to achieve sustainable water production. Renewable energies or waste heat-powered MD are well studied in lab-scale, but extension and optimization for large-scale remain essential. Low-grade heat coming from solar energy, geothermal energy, and waste heat is suitable for MD system. However, the location where the MD process would be implemented is a primordial factor for successful operation.

3. Reduce membrane fouling tendency and pore-clogging in order to extend the membrane lifespan. Membrane fouling is a dynamic process that is affected by various factors including feed water quality, membrane properties, and operating conditions. Develop an in-depth understanding of fouling mechanism to make a prevention strategy to mitigate membrane fouling and improve the performance of used membrane.

Author details

Ali Boubakri[1*], Salah Al-Tahar Bouguecha[2] and Amor Hafiane[1]

1 Laboratory Water, Membranes and Environmental Biotechnology, Center of Water Research and Technologies (CERTE), Soliman, Tunisia

2 Faculty of Engineering, Department of Mechanical Engineering, King Abdul-Aziz University, Jeddah, Saudi Arabia

*Address all correspondence to: ali.boubakri@certe.rnrt.tn

References

[1] Ashoor BB, Mansour S, Giwa A, Dufour V, Hasan SW. Principles and applications of direct contact membrane distillation (DCMD): a comprehensive review. Desalination. 2016;**398**:222-246. DOI: 10.1016/j.desal.2016.07.043

[2] Skuse C, Gallego-Schmid A, Azapagic A, Gorgojo P. Can emerging membrane-based desalination technologies replace reverse osmosis? Desalination. 2021;**500**:114844. DOI: 10.1016/j.desal.2020.114844

[3] Bodell BR. Distillation of saline water using silicone-rubber membrane, United States Patent No. 3,361,645, 1968

[4] Findley ME. Vaporization through porous membranes. Industrial and Engineering Chemistry Process Design and Development. 1967;**6**:226-230. DOI: 10.1021/i260022a013

[5] Boubakri A, Bouguecha SA, Hafiane A. FO – MD integrated process for nitrate removal from contaminated groundwater using seawater as draw solution to supply clean water for rural communities. Separation and Purification Technology. 2022;**298**: 121621. DOI: 10.1016/j.seppur.2022.121621

[6] Ahmed FE, Lalia BS, Hashaikeh R, Hilal N. Alternative heating techniques in membrane distillation: a review. Desalination. 2020;**496**:114713. DOI: 10.1016/j.desal.2020.114713

[7] Shirazi MMA, Kargari A, Ismail AF, Matsuura T. Computational Fluid Dynamic (CFD) opportunities applied to the membrane distillation process: state-of-the-art and perspectives. Desalination. 2016;**377**:73-90. DOI: 10.1016/j.desal.2015.09.010

[8] Alsebaeai MK, Ahmad AL. Membrane distillation: progress in the improvement of dedicated membranes for enhanced hydrophobicity and desalination performance. Journal of Industrial and Engineering Chemistry. 2020;**86**:13-34. DOI: 10.1016/j.jiec.2020.03.006

[9] Saffarini RB, Mansoor B, Thomas R, Arafat HA. Effect of temperature-dependent microstructure evolution on pore wetting in PTFE membranes under membrane distillation conditions. Journal of Membrane Science. 2013;**429**:282-294. DOI: 10.1016/j.memsci.2012.11.049

[10] El-bourawi MS, Ding Z, Ma R, Khayet M. A framework for better understanding membrane distillation separation process. Journal of Membrane Science. 2006;**285**:4-29. DOI: 10.1016/j.memsci.2006.08.002

[11] Boubakri A, Bouguecha SAT, Hafiane A. Box–Behnken design assisted by theoretical mass and heat transfer using for multi-responses optimization of membrane distillation process. Chemical Papers. 2021;**75**:6009-6024. DOI: 10.1007/s11696-021-01778-6

[12] Boubakri A, Elgharbi S, Bouguecha SA-T, Hafiane A. Energetic performance and permeate flux investigation of direct-contact membrane distillation for seawater desalination. Chemical Engineering and Technology. 2020;**43**:2457-2468. DOI: 10.1002/ceat.201900425

[13] Subramani A, Jacangelo JG. Treatment technologies for reverse osmosis concentrate volume minimization: a review. Separation and Purification Technology. 2014;**122**:472-489. DOI: 10.1016/j.seppur.2013.12.004

[14] Drioli E, Ali A, Macedonio F. Membrane distillation: recent

developments and perspectives. Desalination. 2015;**356**:56-84. DOI: 10.1016/j.desal.2014.10.028

[15] Saffarini RB, Summers EK, Arafat HA, J.H. Lienhard V. Technical evaluation of stand-alone solar powered membrane distillation systems. Desalination. 2012;**286**:332-341. DOI: 10.1016/j.desal.2011.11.044

[16] El Mokhtar I, Boubakri A, Bouguecha SA-T, Hafiane A. Modeling and experimental study of air gap membrane distillation unit: application for seawater desalination. Desalination and Water Treatment. 2019;**154**:72-81. DOI: 10.5004/dwt.2019.23889

[17] Usman HS, Touati K, Rahaman MS. An economic evaluation of renewable energy-powered membrane distillation for desalination of brackish water. Renewable Energy. 2021;**169**: 1294-1304. DOI: 10.1016/j.renene.2021.01.087

[18] Lou XY, Ji ZG, Xu Z, Bai AP, Resina-Gallego M. Separation and recycling of concentrated heavy metal wastewater by tube membrane distillation integrated with crystallization. Membranes (Basel). 2020;**10**(1):19. DOI: 10.3390/membranes10010019

[19] Criscuoli A, Drioli E. Date juice concentration by vacuum membrane distillation. Separation and Purification Technology. 2020;**251**:117301. DOI: 10.1016/j.seppur.2020.117301

[20] Boubakri A, Elgharbi S, Dhaouadi I, Mansour D, Al-Tahar Bouguecha S. Optimization and prediction of lead removal from aqueous solution using FO–MD hybrid process: statistical and artificial intelligence analysis. Journal of Environmental Management. 2023;**337**:117731. DOI: 10.1016/j.jenvman.2023.117731

[21] Naidu G, Tijing L, Johir MAH, Shon H, Vigneswaran S. Hybrid membrane distillation: resource, nutrient and energy recovery. Journal of Membrane Science. 2020;**599**:117832. DOI: 10.1016/j.memsci.2020.117832

[22] Ngo MTT, Bui XT, Vo TKQ, Doan PVM, Nguyen HNM, Nguyen TH, et al. Mitigation of thermal energy in membrane distillation for environmental sustainability. Current Pollution Reports. 2023;**53**:13506-13513. DOI: 10.1007/s40726-023-00249-8

[23] Jia X, Lan L, Zhang X, Wang T, Wang Y, Ye C, et al. Pilot-scale vacuum membrane distillation for decontamination of simulated radioactive wastewater: System design and performance evaluation. Separation and Purification Technology. 2021;**275**:119129. DOI: 10.1016/j.seppur.2021.119129

[24] Meindersma GW, Guijt CM, de Haan AB. Desalination and water recycling by air gap membrane distillation. Desalination. 2006;**187**:291-301. DOI: 10.1016/j.desal.2005.04.088

[25] Alkhudhiri A, Darwish N, Hilal N. Membrane distillation: a comprehensive review. Desalination. 2012;**287**:2-18. DOI: 10.1016/j.desal.2011.08.027

[26] Lee WJ, Ng ZC, Hubadillah SK, Goh PS, Lau WJ, Othman MHD, et al. Fouling mitigation in forward osmosis and membrane distillation for desalination. Desalination. 2020;**480**:114338. DOI: 10.1016/j.desal.2020.114338

[27] Rajwade K, Barrios AC, Garcia-Segura S, Perreault F. Pore wetting in membrane distillation treatment of municipal wastewater desalination brine and its mitigation by foam fractionation. Chemosphere. 2020;**257**:127214. DOI: 10.1016/j.chemosphere.2020.127214

Chapter 8

Design, Simulation, and Comparative Analysis of a Carbonating Tower

Jaime Alfonzo Irahola Ferreira

Abstract

In many books, the Solvay process is mentioned as one of the most widely used processes for the production of sodium carbonate. However, on the contrary, no information has been found in the literature that allows for the availability of the functionality of thermodynamic, kinetic, physical, and physicochemical properties with temperature for the exothermic reactions that occur. As it is known, this type of reaction implies that the tower should be cooled. However, no information or heuristic has been found to assign or select the optimal stage for cooling. In this work, a carbonating tower is designed using a simulator to consider the aforementioned functionality, solely based on the input stream data. The beneficial effect of cooling on column performance is confirmed. Subsequently, a cooling mode is established, and the optimal stage to be cooled is determined. Additionally, a heuristic has been proposed to choose the best candidate stages to be cooled. Finally, a sensitivity study of sodium bicarbonate performance is conducted; the variation of concentration and temperature of the product stream is examined in response to changes in the temperature of the feed streams, and a nomogram is presented to solve the proposed study.

Keywords: design, simulation, comparative analysis, carbonating tower, Solvay soda

1. Introduction

The final product of a Solvay soda plant is sodium carbonate (Na_2CO_3). It is also known as trona, natron, barrilla, soda ash, Solvay soda, among other names. It can be found in nature or produced artificially. Sodium carbonate is used as a raw material for the saponification of fatty acids in the manufacture of soaps and detergents. It is also used as a flux in glass furnaces to produce glass containers, flat glass, insulation fibers, and glassware. As a source of alkalinity and sodium ions, it is used in the production of chemicals such as chromium compounds, pigments, and sodium bicarbonate. It is widely used in other industries and processes, including the paper industry, textile industry, metallurgy, mining, oil and gas, water treatment, food processing, etc. [1]. In addition to its traditional uses, it should be noted that sodium carbonate is an important component in the production of lithium carbonate. Briefly, after two previous stages of precipitation and purification of lithium-containing brine,

 IntechOpen

in a final stage, lithium is precipitated as lithium carbonate from the purified brine by the addition of sodium carbonate [2].

According to the report [3], the "lithium triangle," which is the geographical area located in South America, at the border of Argentina, Bolivia, and Chile, concentrates over 85% of the world's reserves of this soft metal. Within the triangle, there are the salt flats of Uyuni (Bolivia); Olaroz-Cauchari, Salinas Grandes, Rincón, and Hombre Muerto (Argentina); and Atacama (Chile), among the largest ones.

In the work [4], there is an extensive development regarding the design and specification of all the equipment involved in the Solvay soda process. They consider a carbonating tower with a cooling stage but do not take into account the temperature functionality of all the properties in each stage of the tower. In the bibliography [5], a study can also be seen regarding the improved manufacturing of soda ash using the AspenPlus simulator. While the simulation of a real plant is presented, little to nothing is mentioned about the study of each stage. However, it is valuable to know that modeling with AspenPlus resulted in highly representative simulations of reality.

In many textbooks, the Solvay process is mentioned as one of the most widely used processes. However, on the contrary, no information has been found in the literature that allows for the availability of the temperature functionality of thermodynamic, kinetic, physical, and physicochemical properties. In that sense, the simulator used is the tool that allows us to overcome such drawbacks and conduct design and/or analysis studies.

This chapter deals with the design of a carbonating tower for the production of sodium bicarbonate. The model to be simulated is built based on the equipment models available in the simulator. In order to find a new design and considering a tower of 13 perforated plates, the only data taken into account are the flow rates and properties of the inlet streams. Once the tower is found, studies are conducted to determine whether or not to cool the column. Upon obtaining an affirmative response, the next step is to determine the best stage to be cooled. Based on the results obtained, a heuristic is proposed that considers the worst and best candidate stages to be cooled. Finally, taking into account the best tower found (tower with the cooled tray), the impact on the sodium bicarbonate yield due to variations in the temperature of the inlet streams is studied.

1.1 Solvay process

The discovery of the chemistry of the ammonia-soda process dates back to the early 1800s. Some British and French plants operated in 1840–1860, but without success. The aforementioned process is generally called the Solvay process because in 1865 Ernest Solvay started the first truly successful plant in Couillet, Belgium [6]. Then, in 1874, the first successful ammonia and soda plant was erected in England. The ammonia-soda process is the dominant technology used worldwide, which is why this process is selected for Solvay soda production. In this process, basically three stages are distinguished: the absorption stage, the carbonating stage and the ammonia recovery stage. The overall reaction is:

$$CaCO_3 + 2NaCl \rightarrow Na_2CO_3 + CaCl_2$$

However, this reaction is not carried out directly. The Solvay process uses an intermediate step. $(NH_4)HCO_3$ is formed to obtain Na_2CO_3 from NaCl and $CaCO_3$.

The necessary ammonia (NH_3) is recycled. The reactive substances are calcium carbonate (main CO_2 generator), sodium chloride, and ammonia (intermediate component).

In the carbonating stage, more specifically, in the tower used for the production of sodium bicarbonate, the following reactions basically occur [7].

$$NH_3 + H_2O + CO_2 \rightarrow (NH_4)HCO_3$$
$$(NH_4)HCO_3 + NaCl \rightarrow NH_4Cl + NaHCO_3$$

The carbon dioxide used is obtained from the calcination of limestone and recovered during the decomposition of bicarbonate into sodium carbonate. Ammonia is introduced into the tower as a brine solution previously obtained in an absorption tower.

The carbonating tower acts as the heart of the Solvay process. Sodium bicarbonate is formed by the absorption of carbon dioxide in the ammoniacal brine. It is then washed, filtered, dried, cooled, and finally sieved. Filtration is carried out in a rotary filter, and for the drying or calcination stage, a steam-heated calciner is generally used, where the bicarbonate decomposes into sodium carbonate according to the reaction:

$$2NaHCO_3 \rightarrow Na_2CO_3 + CO_2 + H_2O$$

2. Carbonating tower design

2.1 Considerations

The absorption of carbon dioxide takes place with chemical reaction, and since the reactions that occur are highly exothermic, a cooling system is generally required for the column. On the other hand, while it is true that the absorption of a gas can be carried out in a packed or tray tower, in this case, only a tray tower can be considered since it would be more complicated to install the cooling circuit in a packed column, as well as extracting side streams for cooling purposes.

The product's outlet temperature is adopted according to the report [8], which establishes that the ideal temperature is between 25°C and 30°C. Lower temperatures can cause salt or ammonium bicarbonate precipitation, while higher temperatures hinder the complete precipitation of the formed bicarbonate.

2.2 Hypothetical problem to be solved

The aim is to design a carbonating tower for which only the flow rates and compositions of the two inlet streams are known at the top and bottom. Additionally, the operating conditions, flow rates, and compositions of the outlet streams at the top (GASO) and bottom (BICARBO), as well as all physical design parameters of the column, hydraulic conditions of all trays, and installation costs, need to be calculated (**Figure 1**).

2.2.1 Data

A tower with 13 sieve trays is considered. **Figure 1** shows the schematic of the simulated carbonating column. The ammoniacal brine stream (SALMU) enters at

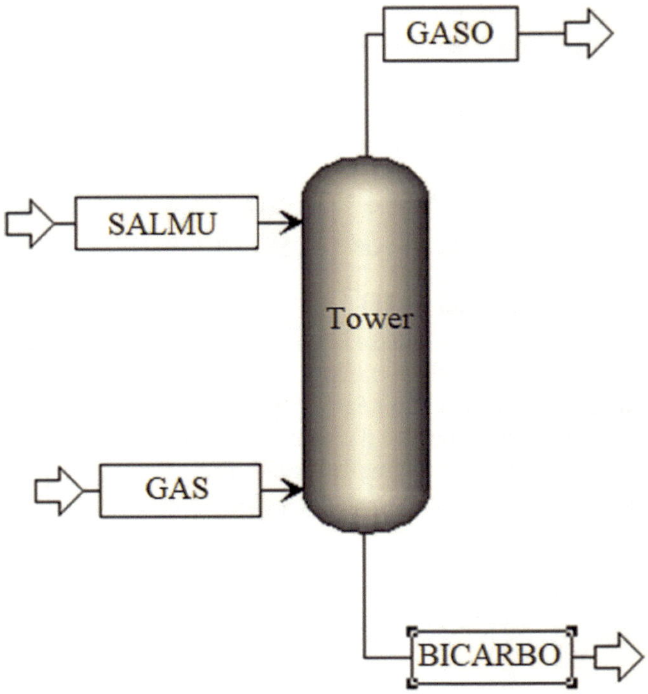

Figure 1.
Carbonating tower.

Stream	GAS		SALMU
Mass Flow Kg/h	70,000		200,874
T [°C′]	60		25
Pressure Bar	2.5		1.0125
Mass Fraction		Mass Fraction	
CO_2	0.460274	NaCl	0.228948
CO	0.00468233	NH_4OH	0.186464
N_2	0.506162	H_2O	0.584587
O_2	0.0288817		

Table 1.
Input streams data.

Tray 1, while the gas stream (GAS) enters at Tray 13. Flow rate, pressure, temperature, and composition information is provided in **Table 1**. The stream compositions have been obtained from [4].

Assumption: a) The presence of a chemical reaction and the probable formation of sodium bicarbonate (solid) are considered based on the operating temperature of the trays.

Restriction: The temperature of the outlet stream, mainly containing sodium bicarbonate, must be within the range of [25, 30]°C.

2.3 Resolution methodology

To solve the problem, it is modeled and simulated using the AspenPlus simulator [9]. The calculation method for physicochemical properties is selected, the components are loaded, and input data for the inlet streams are entered. Then, runs are performed to verify the correct operation of the simulator and to ensure that the reactions proposed by the simulator align with those proposed in the literature (**Table 2**). With the correct model, initially, a simulation is performed without considering the internal column calculation. The desired results are obtained, and then the mentioned internal column design is added. It is verified that all trays operate correctly, costs are calculated, and the simulation is concluded.

2.4 Results

Table 3 presents the overall mass and energy balance as reported by the simulator. It is correct because the mass entering the tower is equal to the mass exiting. The simulator works with a precision greater than 10^{-6}, so if the introduced mole fractions do not have equal or higher precision, small errors occur in the "Relative difference" column. **Table 4** shows a portion of the information obtained after the simulation. It includes values of mass flow rate, density, enthalpy, average molecular weight of the stream, phase state, and operating conditions for each stream and component. Analyzing the top outlet stream (GASO), it is verified that no component is present in the gas phase that, a priori, is known not to exist under the operating conditions of the tower.

Reaction	Type	Stoichiometry
1	Equilibrium	$NH_3 + H_2O \leftrightarrow OH^- + NH_4^+$
2	Equilibrium	$H_2O + HCO_3^- \leftrightarrow CO_3^{2-} + H_3O^+$
3	Equilibrium	$2\ H_2O + CO_2 \leftrightarrow HCO_3^- + H_3O^+$
4	Equilibrium	$2\ H_2O \leftrightarrow OH^- + H_3O^+$
$NaHCO_3$	Salt	$NaHCO_3 \leftrightarrow HCO_3^- + Na^+$
NH_4HCO_3	Dissociation	$NH_4HCO_3 \rightarrow HCO_3^- + NH_4^+$
NH_4CL	Dissociation	$NH_4CL \rightarrow Cl^- + NH_4^+$

Table 2.
Chemical model.

Total	Units	In	Out	Generated	Relative difference
Mole flow	kmol/hr	12,301,416	11,825,826	−299,2414	0,01433567
Mass flow	kg/hr	270,874,49	270,874,49		−3,74E-13
Enthalpy	MMkcal/hr	−680,1975	−680,1976		3,44E-08

Table 3.
Global mass and energy balance in the column.

Stream name	**Units**	**GAS**	**SALMU**	**BICARBO**	**GASO**
Stream Class		CONVEN	CONVEN	CONVEN	CONVEN
MIXED Substream					
Phase		Vapor Phase	Liquid Phase		Vapor Phase
Temperature	C	60	25	25.3616	47.7403
Pressure	Bar	2.5	1.01253	1.1	1
Molar Vapor Fraction		1	0	0	1
Molar Liquid Fraction		0	1	0.980726	0
Molar Solid Fraction		0	0	0.0192741	0
Mass Vapor Fraction		1	0	0	1
Mass Liquid Fraction		0	1	0.925559	0
Mass Solid Fraction		0	0	0.0744415	0
Average MW		33.7875	19.6365	21.7507	26.7028
Mass Flows	kg/hr	70,000	200,874	197,248	73,626
H_2O	kg/hr	0	136,621	126,910	4352.8
CO_2	kg/hr	32219.2	0	85.4751	18895.5
CO	kg/hr	327.763	0	2.77842	324.985
N_2	kg/hr	35431.4	0	252.067	35179.3
O_2	kg/hr	2021.72	0	38.7442	1982.98
Na^+	kg/hr	0	18088.5	14070.3	0
CL-	kg/hr	0	27900.9	27900.9	0
NH_4+	kg/hr	0	63.8135	5524.43	0
OH-	kg/hr	0	5.76E+01	3.84E-03	0
H_3O+	kg/hr	0	4.92E-09	7.61E-05	0
NH_3	kg/hr	0	18142.8	96.67	12890.5
$NaHCO_3$	kg/hr	0	0	14683.5	0
NH4HCO3	kg/hr	0	0	0	0
NH4CL(S)	kg/hr	0	0	0	0
HCO_3–	kg/hr	0	0	7364.75	0
CO_3–	kg/hr	0	0	318.918	0

Table 4.
Global information of the carbonating tower designed.

When examining the results of the BICARBO stream, it is found that:

- The simulator not only presents it in the liquid phase but also reports that a liquid and solid phase coexist, as expected based on the outlet temperature.

- It exits at 25.4°C, therefore, the sodium bicarbonate exits at that temperature, and consequently, the required outlet temperature range of [25,30]°C is satisfied.

- It is composed of a liquid mass fraction (92.5%) and a solid mass fraction (7.5%). The mass fraction of the solid phase shows that it is composed solely of $NaHCO_3$ (14683.5 kg/h).
- 92.9% of the liquid water that entered has passed into this stream.
- No ammonium chloride or ammonium bicarbonate is formed.
- The inert components do not remain completely unchanged, but instead react in the liquid phase in proportion to their inlet composition: 0.8% CO, 0.7% N_2, and 1.9% O_2.
- Only 42% of the CO_2 is utilized.

Based on the above, we can conclude that the tower is operating correctly. However, the utilization of CO_2 could be improved by achieving a better flow rate ratio and maintaining correct hydraulic operating conditions.

The data regarding column sizing and the most important operating conditions are presented in **Table 5**. It should be noted that the spacing between trays is in accordance with what is observed in the literature (0.6 m). The trays function well, hydraulically, since none of them show weeping and a typical allowed value of 80% flooding is also respected.

Moreover, a cost table is also reported (**Table 6**) in which the total capital cost, operating cost and equipment cost (calculated by the simulator), among others, are highlighted. According to the above, it can be seen that a complete and detailed design of the carbonating column has been carried out knowing only the inlet streams.

3. Impact of cooling on the performance of a carbonating column

At this point, the newly designed tower will be considered with an additional cooling stage. The cooling will be done as follows: part of the liquid will be extracted

Section starting stage	1
Section ending stage	13
Calculation mode	Sizing
Tray type	Sieve
Number of passes	1
Tray spacing [m]	0.6096
Section diameter [m]	3.27398
Section height [m]	7.9248
Section pressure drop [bar]	0.1517
Section head loss (Hot liquid height) [m]	1.30429
Trays with weeping	None
Maximum % jet flood	80.0006

Table 5.
Carbonating column design data.

Total capital cost [USD]	2,627,650
Total Operating Cost [USD/Year]	999,672
Total Raw Materials Cost [USD/Year]	0
Total Product Sales [USD/Year]	0
Total Utilities Cost [USD/Year]	35544.4
P.O. Period [Year]	0
Equipment Cost [USD]	254,500
Total Installed Cost [USD]	256,100

Table 6.
Costs of the designed column.

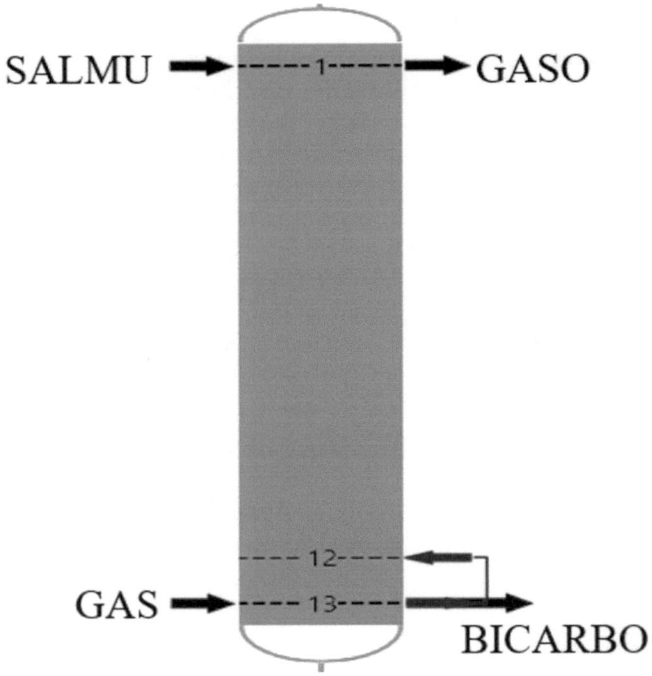

Figure 2.
Carbonating column, streams, and cooling scheme.

from stage 13, this stream will be cooled indirectly by an external heat exchanger and then re-enter stage 12 (**Figure 2**). The data, assumptions, and constraints are the same as those considered for the previous tower, which did not consider cooling. The problem is solved by adding the cooling condition to the model of the designed carbonating column.

3.1 Discussion and results

The basic sizing achieved is common to both columns and is presented in **Table 5**. The simulation results for the column with cooling are shown in **Table 7**. Both

Stream name	Units	GAS	SALMU	BICARBO	GASO
Stream Class		CONVEN	CONVEN	CONVEN	CONVEN
MIXED Substream					
Phase		Vapor Phase	Liquid Phase		Vapor Phase
Temperature	C	60	25	27.0485	50.026
Pressure	Bar	2.5	2	1	1
Molar Vapor Fraction		1	0	0	1
Molar Liquid Fraction		0	1	0.9635029	0
Molar Solid Fraction		0	0	0.0364971	0
Mass Vapor Fraction		1	0	0	1
Mass Liquid Fraction		0	1	0.8638933	0
Mass Solid Fraction		0	0	0.1361067	0
Average MW		33.7875	19.6365	22.5265	27.176
Flujo Molar	kmol/hr	2.071.77	10.229.65	9.313.33	2.247.48
H_2O	kmol/hr	0	7.586.28	6.998.02	188.2803
CO_2	kmol/hr	732.0907	0	8.4931	322.9076
CO	kmol/hr	11.7015	0	0.891	10.8105
N_2	kmol/hr	1.264.80	0	81.0838	1.183.71
O_2	kmol/hr	63.1811	0	10.7223	52.4588
Na^+	kmol/hr	0	786.827	446.917	0
CL-	kmol/hr	0	786.977	786.977	0
NH_4+	kmol/hr	0	0.87	401.796	0
OH-	kmol/hr	0	0.719	0	0
H_3O+	kmol/hr	0	0	0	0
NH_3	kmol/hr	0	1.067.98	177.736	489.315
$NaHCO_3$	kmol/hr	0	0	339.91	0
NH_4HCO_3	kmol/hr	0	0	0	0
$NH_4CL(S)$	kmol/hr	0	0	0	0
HCO_3–	kmol/hr	0	0	59.825	0
CO_3–	kmol/hr	0	0	0.956	0

Table 7.
Overall information on carbonating tower with cooling.

designs, with and without cooling, satisfy the requirement the outlet temperature of the liquid stream is within the desired range of [25–30]°C. Thus, the temperatures are 27.05°C and 25.36°C for the columns with and without cooling, respectively.

The total capital costs of the column with cooling and without cooling are correspondingly: USD 2,972,670 and USD 2,627,650, that is. the former is 13.1% higher. However, this difference decreases when comparing the operating costs, which are USD 1,029,020 per year for the column with cooling and USD 999,672 per year for the column without cooling. The column with cooling is only 2.9% more expensive, a value that could potentially be reduced or even eliminated with changes in column operation.

If the stoichiometry of the reactions involved is considered, it is found that one mole of carbon dioxide produces one mole of sodium bicarbonate. Therefore, by observing **Tables 4** and 7, it can be seen that without cooling, 174.789 kmol/hr. of sodium bicarbonate is produced, while with cooling, 339.91 kmol/hr. is produced, representing a 94.5% increase with the latter. Furthermore, the amount of solid sodium bicarbonate produced in the cooled column is almost double compared to the other column, specifically 82.8% more. This information is interesting as the simulator reports that it is the only species in solid phase.

When analyzing the unreacted CO_2 exiting the column, it is observed that the column without cooling loses 33% more CO_2 compared to the column with cooling (**Tables 4** and 7), which somehow corroborates the sodium bicarbonate production data.

Finally, it is interesting to examine the temperature profiles for both columns (**Figure 3**). In graph (a), both curves are monotonically increasing, with the outlet temperature of Tray 13 being lower than that of the other column in graph (b). In this graph, Trays 12 and 13 operate at practically the same temperature, which is the lowest throughout the column, namely 26.98°C and 27.05°C, respectively. This

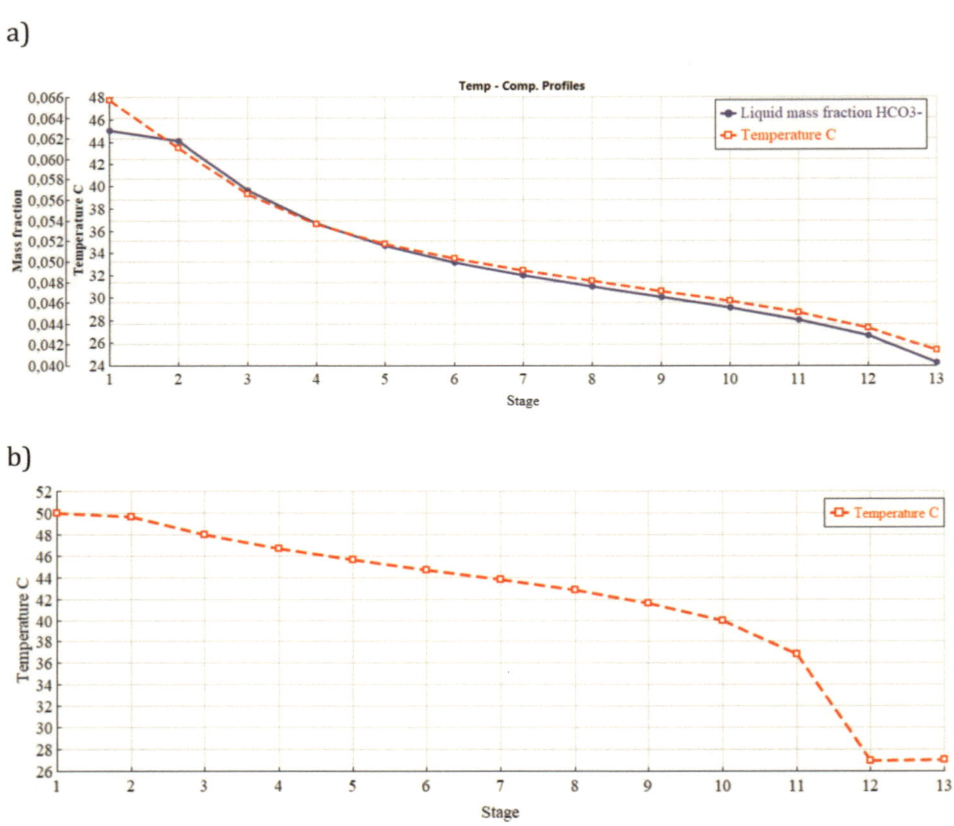

Figure 3.
Temperature and composition profile of the carbonating tower. (a) Tower without cooling and (b) tower with cooling.

explains the higher amount of solid-phase sodium bicarbonate compared to the liquid phase and, as mentioned earlier, in a greater proportion than in the column without cooling.

4. Optimal stage to cool in a carbonating column

As mentioned earlier, highly exothermic reactions occur in the carbonating column, especially due to CO_2 absorption. Therefore, it can be assumed that cooling the column is necessary, and indeed, as proven in the previous section, it results in improved sodium bicarbonate production.

Now, the goal is to determine, which stage of the carbonating column should be cooled to achieve the highest yield in $NaHCO_3$ production. All the data considered in the cooled column (**Table** 7) are adopted [10].

It should be noted that for the column design, only the flow rates and compositions of the two feed streams to the tower are considered. The liquid stream consisting of ammoniated brine (SALMU) previously obtained in an absorption tower within the same plant, enters Tray 1. The gas stream (GAS), obtained from the calcination of limestone and recovered during the decomposition of bicarbonate a).

into sodium carbonate in a Solvay soda plant, enters Tray 13. The product, $NaHCO_3$, is obtained in the liquid outlet stream (BICARBO), while the unreacted gases exit through the gas outlet stream (GASO).

4.1 Methodology for modeling cooling and finding the optimal stage

A 13-stage column is initially designed without cooling. Then, the condition of cooling in the last stage is added, and the problem is solved. The cooling process involves extracting a portion of the liquid from a stage, cooling it through an external heat exchanger, and reintroducing it to the same stage (**Figure 4**).

Next, the amount of heat extracted is varied to achieve maximum $NaHCO_3$ production. Once this point is reached, all the design data is kept as parameters, and the same amount of heat is extracted from each stage of the column. In other words, a simulation is performed for each stage that is cooled. If all the variables are kept fixed and only the tray to be cooled is changed, the study will allow comparing the $NaHCO_3$ yield in each case and therefore choosing the best tray to be cooled.

4.2 Discussion and results

The search for the optimal stage was conducted by cooling from the first stage (Tray 1) to the last stage (Tray 13) in order to establish the finding of a global optimum. The graphical comparison of the mass flow rate of $NaHCO_3$ produced, based on the stage being cooled (**Figure 5**), reveals that when cooling Tray 11, the highest performance of the carbonating column is achieved: 28,584.93 kg/hr. of sodium bicarbonate. However, the difference with the adjacent trays is small. It is only 0.020% and 0.054% more efficient compared to cooling Tray 12 or Tray 10, respectively. Therefore, for practical purposes, either of these trays could be chosen without causing significant differences.

It is interesting to note that as the tray closest to the bottom of the column is cooled, the BICARBO temperature decreases but the GASO temperature increases (**Figure 6**). This is logical since the cooling stage is farther away from GASO and

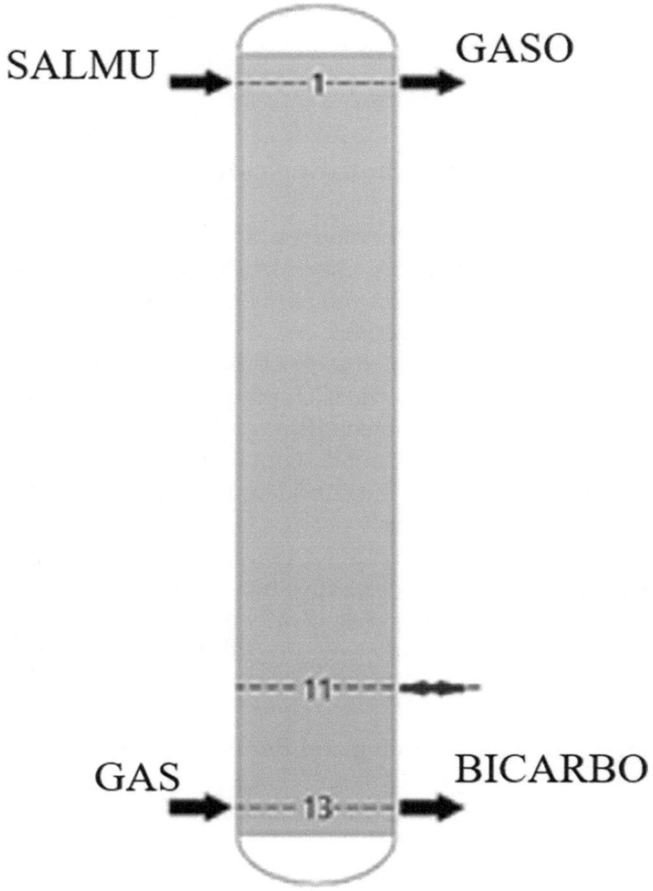

Figure 4.
Scheme of the stage to be cooled in the carbonating column.

closer to BICARBO. Regarding the flow rates of the outlet streams, a similar behavior is observed. The flow rate of BICARBO decreases when cooling is applied to the trays near the column's bottom, while the flow rate of GASO increases (**Figure** 7). This is consistent with the temperature profile seen before since, for example, if the GASO temperature is higher, this justifies the higher flow rate of that gaseous stream. The temperature of BICARBO is 27.3 °C compared to 27.05°C in the cooled tower and 25.36°C in the first non- cooled carbonating tower. This demonstrates that as the temperature in BICARBO increases, the concentration of sodium bicarbonate in BICARBO and the column's performance also increase.

On the other hand, the fact of keeping all the design variables of the column constant and only evaluating the number of plates to be cooled allows us to affirm that the result achieved is valid and the following heuristic could be proposed for the choice of the best stage to be cooled:

1. Discard the first and the last stage.

2. Discard as candidates all the stages from the second (from the top) to the one located in the middle.

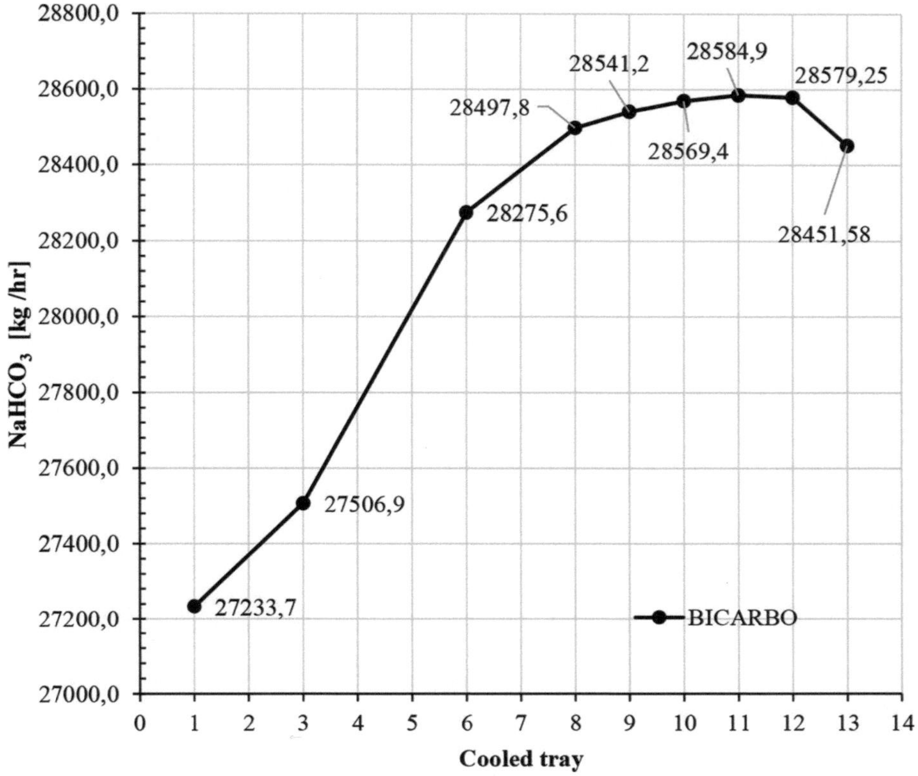

Figure 5.
Flow rate of NaHCO3 according to the cooled tray.

3. The best candidate will be closer to the base of the column than to the middle of the column.

5. Sensitivity of the performance of a carbonating column to changes in temperature of the inlet streams

The objective of this study is to determine the impact on the $NaHCO_3$ performance due to changes in the temperature of the inlet streams. Three cases are studied:

a. The temperature of the liquid feed stream, SALMU, is changed in steps of 5 °C from 10 to 35°C while keeping the rest of the parameters constant.

b. The temperature of the GAS feed stream is changed in steps of 5°C from 30 to 80°C while keeping the rest of the parameters constant.

c. Simultaneous variation of a) and b).

To proceed, we choose the tower with the best performance among the ones studied here. Therefore, we select the column with tray 11 being cooled and conduct

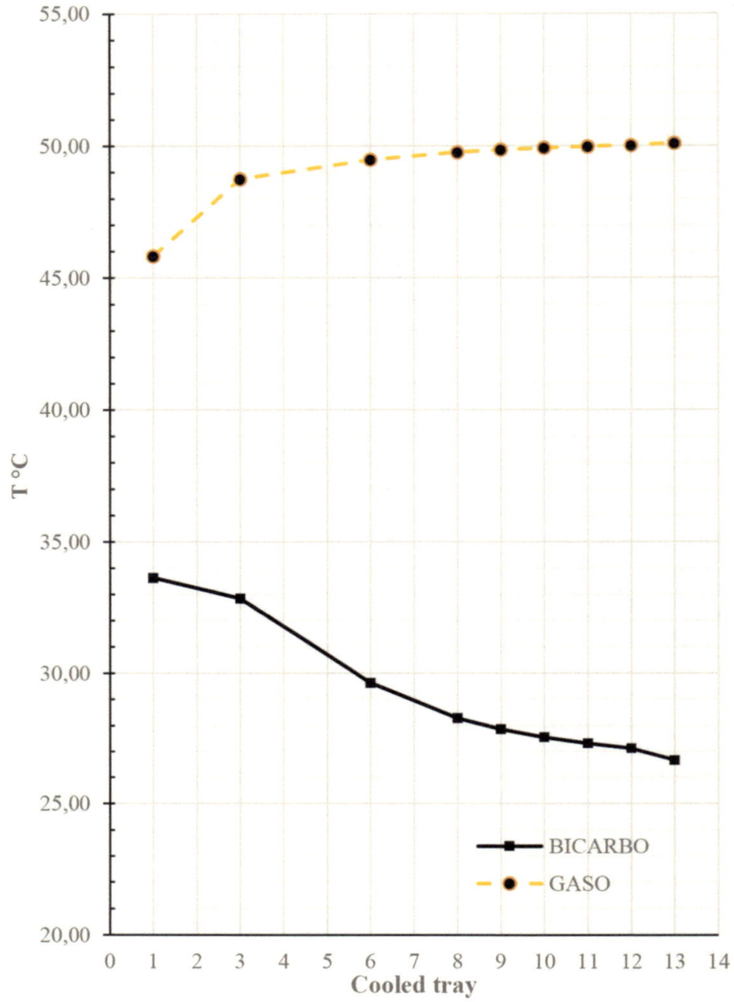

Figure 6.
Temperature of the outlet streams according to the cooled tray.

the sensitivity study. Firstly, we set the temperature of the GAS stream as a parameter and fix it at 30 °C. Then, we simulate for a temperature of 10 °C for the SALMU stream. Next, we move on to the next simulation: keeping the parameter value constant, we increment the temperature of the SALMU stream by a step increment of 5° C, that is. we take the value of 15°C. This process is repeated, taking the same parameter value and incrementing the SALMU stream temperature by the step value (5°C) for each new simulation, until we cover the entire proposed study interval: [10–35]°C (**Figure 8**). Therefore, a total of six simulations will be performed, considering the same parameter value.

Subsequently, the parameter value is increased by a step of 5°C, and the aforementioned procedure is repeated. The parameter increment is done within a range of 30 to 80°C. At the end of the described process, a total of 66 simulations will have been conducted. Similarly, the procedure is followed to address the proposed objectives for case (b) (**Figure 9**) and (c).

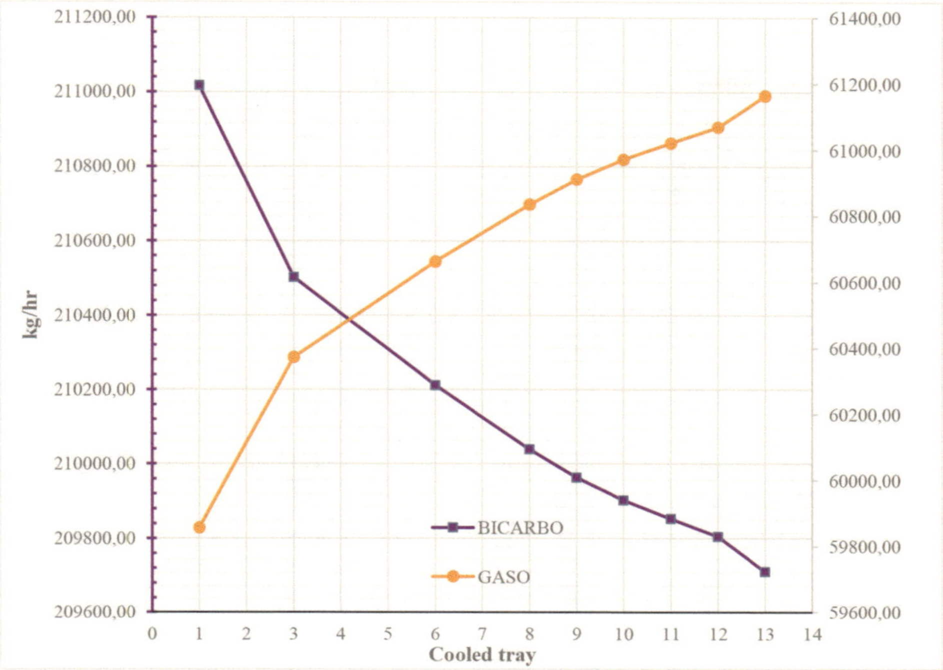

Figure 7.
Mass flow rate of the outlet streams according to the cooled tray.

5.1 Sensitivity analysis results

When analyzing the results for case (a), it can be observed that the production of $NaHCO_3$ decreases as the T_{SALMU} temperature increases (**Figure 8**), for all values of the T_{GAS} parameter. A similar behavior is observed when analyzing case b), where the production of sodium bicarbonate decreases as the T_{GAS} temperature increases, for all values of the T_{SALMU} parameter (**Figure 9**). It is interesting to note that in both cases, the evolution of the curves is nearly linear. Although the curves in **Figures 8** and **9** may appear to be strictly straight and parallel at first glance, they are not strictly parallel. When comparing the slopes, it is evident that they are not parallel. In the best case, that is. the case with the highest concentration (T_{GAS} = 30°C, **Figure 8**), the slope of the curve is smaller than the slopes of the other curves. A similar situation is observed in **Figure 9**.

Furthermore, in the curve corresponding to the most favorable condition mentioned earlier (T_{GAS} = 30°C, **Figure 8**), it can be observed that for every 1°C increase in the T_{SALMU} temperature, the mass fraction of $NaHCO_3$ decreases by $1.0061\ 10^{-3}$, which is equivalent to a flow rate of 259.09 [Kg/h]/°C.

When analyzing the variation in T_{GAS} temperature (**Figure 9**) in the best case (T_{SALMU} = 10°C), it can be observed that the mass fraction of $NaHCO_3$ decreases by.

$1.2003\ 10^{-4}$/°C, or in other words, the $NaHCO_3$ flow rate decreases by 29.342 [Kg/h]/°C. Therefore, the production of the column is more affected by changes in the temperature of the liquid feed stream than the gas stream. Quantitatively, the production of $NaHCO_3$ is eight times more affected by a one-degree celsius variation in the T_{SALMU} stream than in the T_{GAS} stream.

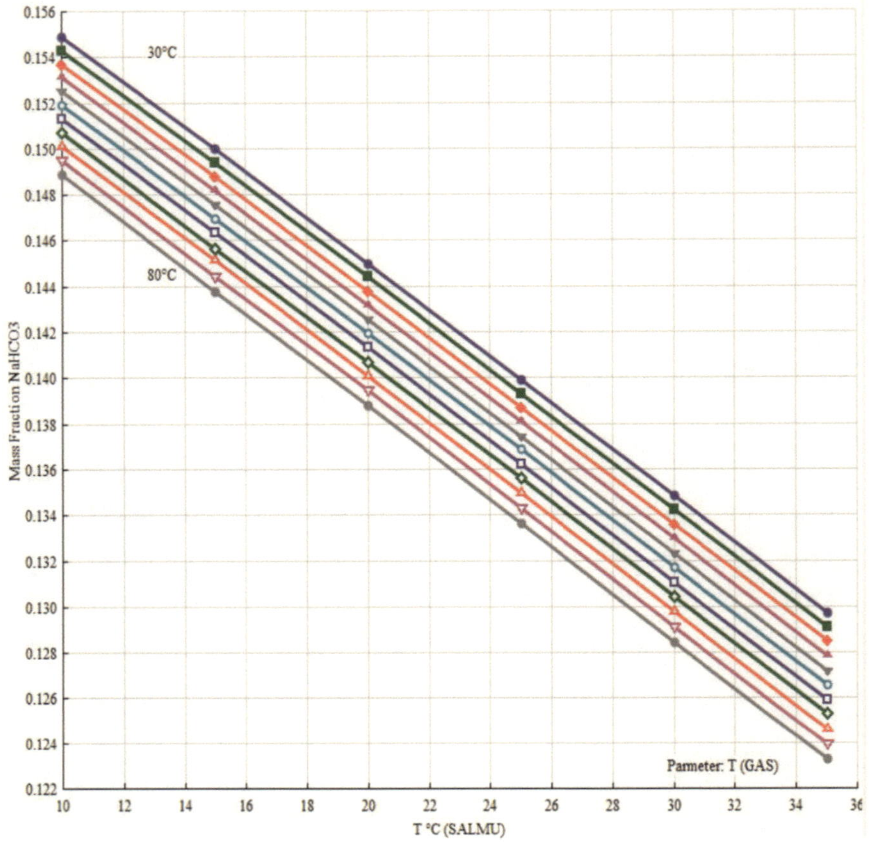

Figure 8.
Evolution of $NaHCO_3$ mass fraction with temperature.

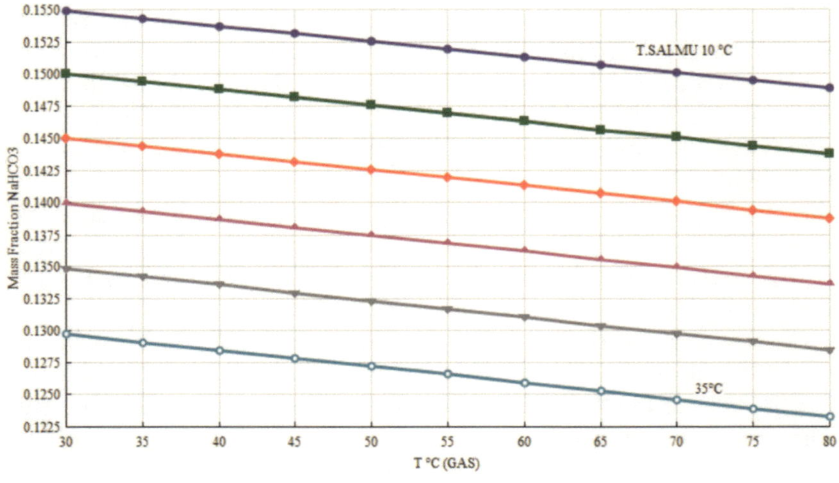

Figure 9.
Relationship between $NaHCO_3$ mass fraction and temperature rise of the gaseous feed stream.

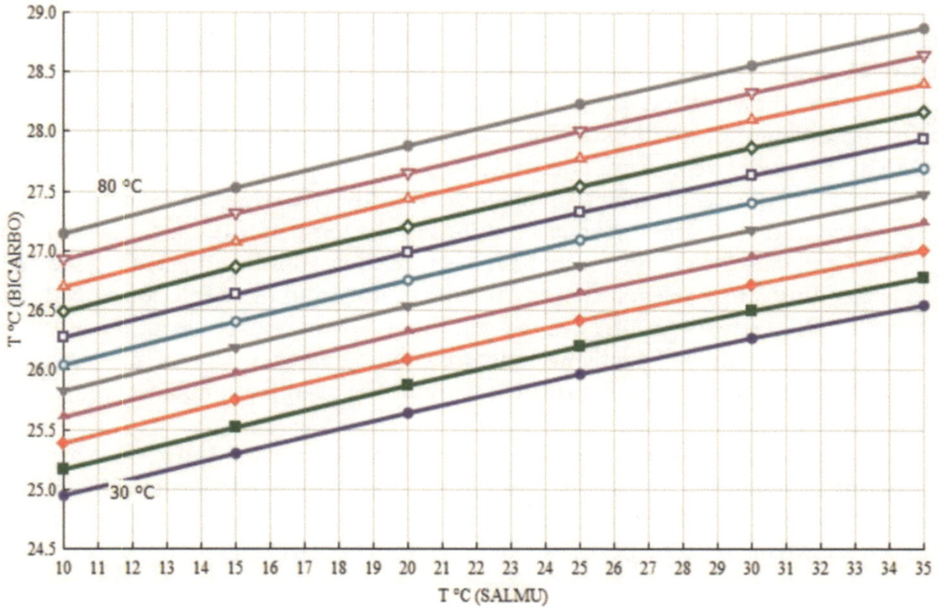

Figure 10.
Relationship between BICARBO stream temperature and liquid feed stream temperature rise.

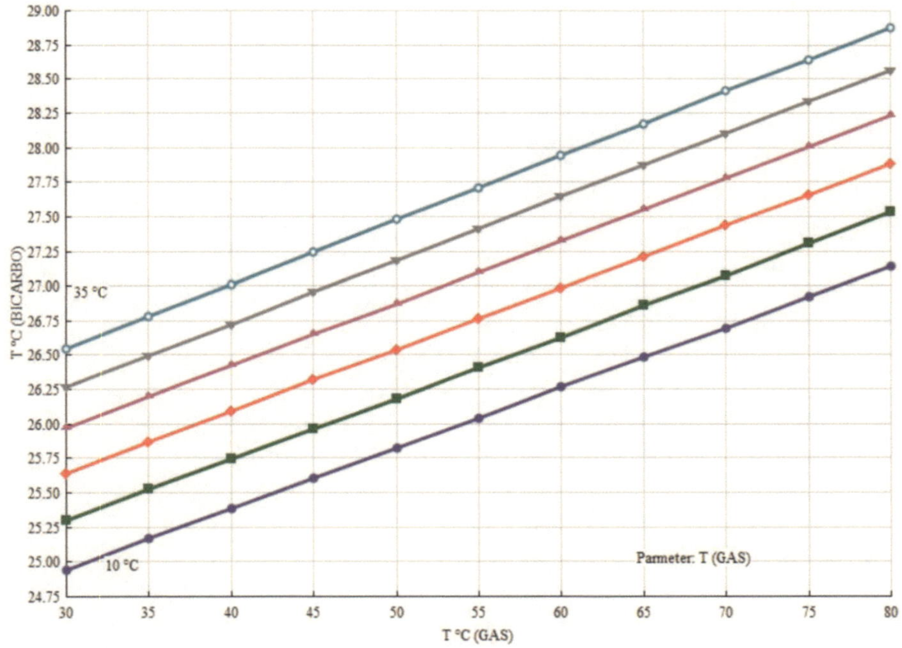

Figure 11.
Relationship between the temperature of the BICARBO stream and the temperature rise of the gaseous feed stream.

Figures 10 and **11** show the impact on the $T_{BICARBO}$ due to the variation in the temperature of the liquid and gas feed streams. Increasing the temperature of these streams results in an increase in the temperature of the product stream in both cases.

Considering the lower curve in **Figure 10**, for every 1°C increase in T_{SALMU}, the temperature $T_{BICARBO}$ increases by only 0.064°C. A similar analysis based on **Figure 11** shows a relationship of 0.044°C T(GAS)/°C T(BICARBO).

Therefore, the variation of the liquid feed stream temperature (T_{SALMU}) impacts 45.6% more than the variation of the gaseous feed stream temperature (T_{GAS}), on the BICARBO product stream temperature ($T_{BICARBO}$).

In the study of case c), which involves the sensitivity analysis of sodium bicarbonate production due to simultaneous variations in the temperatures of the SALMU and GAS feed streams, the results found have been condensed into a graph (**Figure 12**). This graph illustrates the evolution of the temperature and mass fraction of $NaHCO_3$ in the BICARBO stream for any desired condition. Furthermore, the graph serves as a nomogram for the studied problem. Using the nomogram, one can obtain results for any combination of the variables studied without the need for simulation under those specific conditions.

For example, if one wants to determine the temperature and $NaHCO_3$ concentration when T_{SALMU} = 25°C and T_{GAS} = 75°C, they would locate 25°C on the x-axis, follow the mass fraction curves to find the one corresponding to 75°C, and read the approximate value of 0.134 (mass fraction of $NaHCO_3$) on the y-axis.

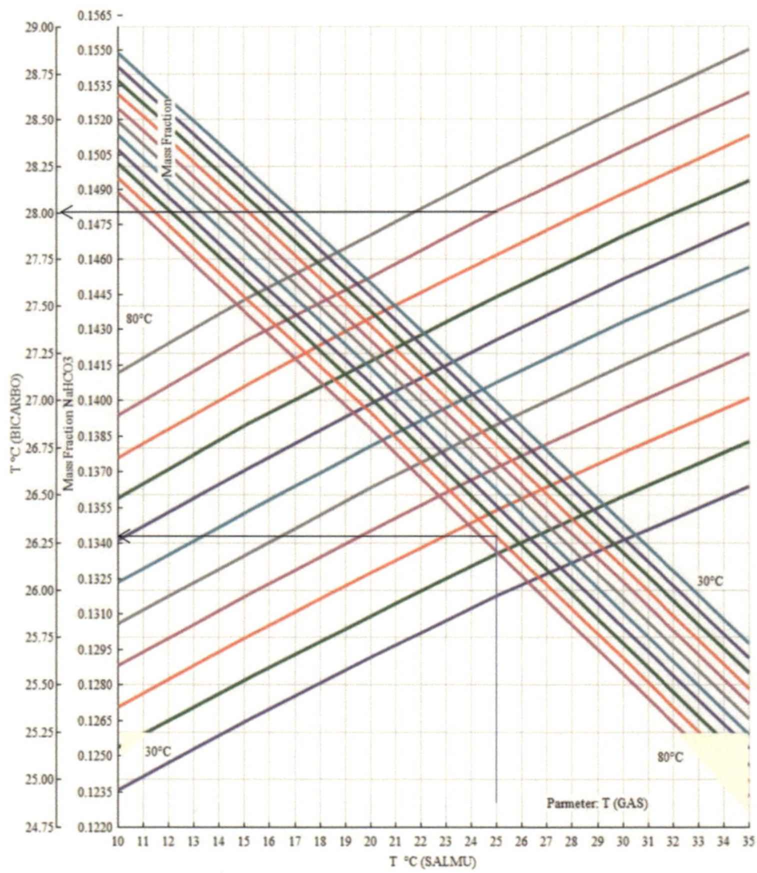

Figure 12.
Simultaneous sensitivity study of temperature and concentration of the BICARBO stream.

Using the temperature curves, $T_{BICARBO}$ can be estimated to be approximately 28°C. (Note: To satisfy operational restrictions, the graph excludes the portion of curves below 25°C.)

6. Conclusions

A carbonation tower has been designed for the production of sodium bicarbonate. The model to be simulated was built based on the equipment models of the simulator. The design is complete and includes the physical dimensioning of a column of sieve trays, the dimensioning of each tray, and the verification of the hydraulic operation of each one of them. It is important to point out that the use of the simulator has taken into account the calculation of the physicochemical and thermodynamic properties as a function of temperature of all the components present, which otherwise would not have been possible due to the scarcity of bibliographic information for this purpose. It also allows a quick calculation of all costs, namely: total capital cost, operating, service, and equipment costs, among others.

It is well-known that the carbonating process is exothermic, and therefore, the column should be cooled. This has been confirmed by the conducted study. It has been found that when a stage of the carbonating column is cooled, the total capital cost of the column is 13% higher than when the column is not cooled. However, the operating costs are similar, with only a 2.9% increase due to cooling. In return, there was a 35.3% improvement in the yield of sodium bicarbonate production.

Once the question of whether or not to cool a carbonating column is satisfied, the next question is how many stages should be cooled and which ones. In this regard, no study has been found in the literature that provides the answer. Therefore, this paper presents the results of the behavior and performance of each carbonating column according to the cooled stage. After analyzing the 13 trays of the studied column, it was found that the optimal stage to be cooled is tray 11. The result of the conducted study is limited by the input data and the cooling process used. However, on the other hand, the fact that all the design variables of the column were kept constant, and only the tray to be cooled was evaluated allows us to assert that the achieved result is valid as a global optimum.

Furthermore, it can be proposed as a heuristic that when choosing the stage to be cooled, it is advisable not to select the stages near the top or middle of the column but rather those closer to the bottom of the column.

From the sensitivity analysis of the $NaHCO_3$ yield in the carbonating column, it has been found that: (i) The concentration of $NaHCO_3$ decreases as the temperature of the feed streams increases and that the influence is eight times greater for variation of the temperature of the feed liquid stream than for variation of the temperature of the gaseous feed stream; (ii)The temperature of the liquid outlet stream (product) increases when the temperature of the feed streams, either one or both of them, increases; and (iii) The impact on the outlet temperature is 45.6% greater per degree celsius of temperature variation in the liquid feed stream compared to the variation in the gaseous feed stream.

Lastly, the sensitivity analysis allowed for the creation of a nomogram, which enables the determination of the conditions of the outlet stream, such as temperature and concentration, based on the temperatures of the inlet streams. Furthermore, it could easily be constructed to display all other variables of the column, including flow rate and concentrations of all species present in both outlet streams.

The findings presented in this study are valid for the given conditions and assumptions. However, the results may vary under different circumstances such as changing the type of column (e.g., having multiple passes or different sections), altering the conditions of the cooling process in the stage, modifying the percentage of flow rate to be cooled, and adjusting the amount of heat exchange, among other factors.

Acknowledgements

Special thanks to IntechOpen for their exceptional collaboration and invaluable support in the publication of this chapter. We would also like to extend our sincere gratitude to Universidad Nacional de Jujuy for their support of scientific research.

Thanks

I would like to thank my parents, my brother and sister John and Liliana, my son "Jaimito-Leo" and my wife Mechy, for their immense encouragement and love.

Author details

Jaime Alfonzo Irahola Ferreira
National University of Jujuy, Jujuy, Argentina

*Address all correspondence to: irahola.j@gmail.com

References

[1] Dirección Nacional de Promoción y Economía Minera. Subsecretaría de Desarrollo Carbonato de Sodio (Soda Ash): características, usos y demanda. Minero. 2022. Available from: https://www.argentina.gob.ar/sites/default/files/soda_ash_serie_de_estudios_para_el_desarrollo_minero_carbonato_de_sodio_soda_ash_caracteristicas_usos_y_demanda.pdf

[2] Diaz Tapia JD. Procedimiento de extracción de litio para la obtención de carbonato de litio, desde una salmuera o mineral y/o arcilla previamente tratada para estar libre de boro. Patente, WO; 2013

[3] The next low cost lithium producer. In: 2011 Electrical Metals Conference, Tuesday, April 12, 2011. Orocobre Limited. Byron Capital Markets

[4] Cervera Gracia L, Izquierdo L, Labrozzi JP, Moreno M, Ordóñez EO, Vidal R. Planta de producción de carbonato sódico. Tesina de Ingeniería Química. Escuela Técnica Superior de Ingeniería. Chapter 1. 2006. pp. 17-57

[5] Cormos AM, Cormos CC, Agachi P. Making soda ash manufacture more sustainable. A modelling study using ASPEN Plus. In: Plesu V, Agachi PS, editors. 17th European Symposium on Computer Aided Process Engineering – ESCAPE17. Elsevier B.V.; 2007. pp. 1-6

[6] Analysis on the Production of Soda Ash Using Solvay Process. 2014. Available from: https://www.academia.edu/8035384/Soda_Ash_Production

[7] Wagialla KM, Al Mutaz IS, el Dahshan ME. The manufacture of soda ash in the Arabian Gulf. International Journal of Production Economics. 1992; **27**:145-153

[8] Federal Environmental Agency. Final Report. 2001. Available from: https://www.umweltbundesamt.de/en/publikationen/annual-report-2001-federal-environmental-agency

[9] Aspen Plus. Aspen Plus User Guide. 10.2, Aspen Technology, Inc.; 2000. Available from: http://www.aspentech.com

[10] Irahola Ferreira JA. Selección de la etapa óptima a refrigerar en una columna de carbonatación. In: Proceedings of the Encuentro Argentino y Latinoamericano de Ingeniería. CLADI/CADI. Facultad de Ingeniería de la UBA. Buenos Aires; 2021